T0321373

Seeing Invisible:
Advanced Antenna Arrays

RIVER PUBLISHERS SERIES IN COMMUNICATIONS AND NETWORKING

Series Editors

ABBAS JAMALIPOUR
The University of Sydney,
Australia

MARINA RUGGIERI
University of Rome Tor Vergata,
Italy

The "River Publishers Series in Communications and Networking" is a series of comprehensive academic and professional books which focus on communication and network systems. Topics range from the theory and use of systems involving all terminals, computers, and information processors to wired and wireless networks and network layouts, protocols, architectures, and implementations. Also covered are developments stemming from new market demands in systems, products, and technologies such as personal communications services, multimedia systems, enterprise networks, and optical communications.

The series includes research monographs, edited volumes, handbooks and textbooks, providing professionals, researchers, educators, and advanced students in the field with an invaluable insight into the latest research and developments.

Topics included in this series include:-

- Communication theory
- Multimedia systems;
- Network architecture;
- Optical communications;
- Personal communication services;
- Telecoms networks;
- Wifi network protocols.

For a list of other books in this series, visit www.riverpublishers.com

Seeing Invisible:
Advanced Antenna Arrays

Pavlo A. Molchanov
IPD Scientific LLC., USA

River Publishers

Routledge
Taylor & Francis Group
NEW YORK AND LONDON

Published 2024 by River Publishers

River Publishers

Alsbjergvej 10, 9260 Gistrup, Denmark

www.riverpublishers.com

Distributed exclusively by Routledge

605 Third Avenue, New York, NY 10017, USA

4 Park Square, Milton Park, Abingdon, Oxon OX14 4RN

Seeing Invisible: Advanced Antenna Arrays / Pavlo A. Molchanov.

Routledge is an imprint of the Taylor & Francis Group, an informa business

ISBN 978-87-7004-023-5 (hardback)

ISBN 978-87-7004-103-4 (paperback)

ISBN 978-1-003-47655-9 (ebook master)

While every effort is made to provide dependable information, the publisher, authors, and editors cannot be held responsible for any errors or omissions.

To my lovely wife, Lyudmyla

Contents

Preface

The purpose of this book is to provide a comprehensive and accessible explanation of antenna arrays, their advancements, and their limitations. By exploring the principles of antenna arrays and their applications, readers will gain a deeper understanding of their design and potential advancements inspired by nature.

Antennas and antenna arrays play a crucial role in sensory systems and circuits, serving as the primary components for various technologies. Without well-designed sensor systems or antenna arrays, it is challenging for artificial intelligence, computers, vision systems, communication systems, and detection systems to operate effectively. This significance extends beyond technical fields like communication, navigation, and radar to encompass healthcare, medical imaging, and even the design of amateur metal detectors and RC vehicles. In the context of present-day military and geopolitical threats, this book will address the functional requirements of the new generation of antenna arrays for radar and anti-missile systems, equipping readers with an understanding of how these arrays can address such challenges.

The book covers the evolution of antennas and antenna array designs, shedding light on the fundamental principles and advancements associated with nature-inspired, next-generation fly eye antenna arrays and their applications.

While the book is accessible to a wide range of readers and does not require specialized education, it will serve as a valuable resource for engineers and technicians involved in the design and development of all-space communication and navigation systems, drones, and hypersonic missile detection systems. The book particularly emphasizes interference immunity and protection in these applications. Furthermore, it can be employed as a textbook for advanced radar technology courses and seminars, providing a comprehensive understanding of the subject matter.

Foreword

Seeing invisible, seeing concealed or underground objects, and seeing behind the corner or through walls – the human desire to see what is normally hidden or concealed has always been a fascination. Our natural visual range is limited, preventing us from seeing through obstacles such as fog, snow, walls, or the ground. However, radio frequency signals have the ability to penetrate these barriers more effectively. For example, we can use our cell phones and Wi-Fi devices indoors, where the signals can pass through walls. While this allows some level of "seeing through walls," it is limited by the resolution of the images captured, which is constrained by the diffraction limit imposed by classical physics.

This book aims to demonstrate that the ability to see the invisible is indeed possible and explores how advanced antenna arrays can facilitate this task. By leveraging the principles and capabilities of antenna arrays, we can push the boundaries of what is perceivable and expand our vision beyond the limitations of the human eye. The book will delve into the technologies and techniques that enable enhanced sensing, imaging, and perception, offering insights into the exciting world of hidden object detection and imaging through advanced antenna array systems.

This book aims to illustrate the possibility of "seeing the invisible" and explores how advanced antenna arrays can enable this capability. By leveraging the unique properties and functionalities of antenna arrays, it becomes feasible to perceive and detect objects or phenomena that are typically hidden or imperceptible to the human senses.

Through a comprehensive examination of the principles, design considerations, and applications of advanced antenna arrays, this book will provide readers with a deeper understanding of how these arrays can enhance our ability to sense and perceive the unseen. It will delve into various techniques and technologies employed in antenna array systems, including signal processing, beamforming, waveform design, and integration with other sensing modalities.

By showcasing real-world examples and case studies, this book will demonstrate the practical applications of advanced antenna arrays in fields such as radar systems, surveillance, remote sensing, medical imaging, and more. It will also explore emerging trends and future prospects in the field, highlighting the potential for further advancements and breakthroughs in "seeing the invisible" through the continued development of antenna array technologies.

Whether you are an engineer, researcher, student, or enthusiast interested in the fascinating realm of antenna arrays and their applications, this book will serve as a valuable resource to deepen your knowledge and appreciation of the remarkable possibilities that lie within the realm of "seeing the invisible."

List of Figures

xvi   *List of Figures*

List of Tables

List of Abbreviations

ADC	Analog-to-digital converter
AESA	Active electronically scanned array
AOA	Angle of arrival
AOV	Angle of view
BPF	Bandpass filter
BPFn	Multi-band (n) pass filter
CG	Clock generator
CRC	Conditioning receiving chain
CW	Continuous wave
CZT	Cittert–Zernike theorem
DAC	Digital-to-analog converter
dB	Decibel
DBF	Digital beam forming
DOA	Direction of arrival
EMP	Electromagnetic pulse
EO	Explosive object
FEC	Front-end circuit
FFT	Fast Fourier transform
FPGA	Field programmable gate array
GEO	Geo earth orbit
GHz	Giga Hertz
GPR	Ground penetrating radar
HPA	High-power amplifier
I/Q	In-phase/quadrature
IED	Improvised explosive device
IF	Intermediate frequency
ISAR	Inverse synthetic aperture radar
ISR	Intelligence, surveillance, and reconnaissance
LED	Light emitting diode
LEO	Low Earth orbit
LNA	Low noise amplifier
LPF	Lowpass filter

MEO	Middle Earth orbit
MIMO	Multiple-input and multiple-output
ONR	Office of Naval Research
RF	Radio frequency
RMS	Root mean square
RTE	Radio technical equipment
SAR	Synthetic aperture radar
SDR	Software-defined radio
SiP	System-in-package
SMART	Spatially matched adaptive radar tomography
SNR	Signal-to-noise ratio
SW	Switchers
SWAP	Size, weight, and power
TAM	Transmitter antenna module
TC	Transmitting chain
TDOA	Time difference of arrival
TT	transceiver band/service translator
TWA	Traveling wave antenna
UAS	Unmanned aerial system
UHF	Ultra-high frequency
USB	Universal serial bus
UXO	Unexploded ordnance
VERA	Passive reconnaissance and surveillance system
VLEO	Very low Earth orbit

1

Nature-inspired Fly Eye: Multi-beam Staring Antenna Array Concept

1.1 Fly Eye Structure

To compensate for its eye's inability to point at a threat, a fly's eye consists of multiple optical sensors positioned at angular intervals, which give the fly the wide-area visual coverage it needs to detect and avoid threats around it. Each sensor is coupled with a detector and connected by a separate neuron interface to the brain (processor). Figure 1.1 depicts a picture and structure of a fly's eye, along with its functioning. In this structure, there are multiple optical sensors positioned at angular intervals. These sensors capture optical signals and produce corresponding analog signals.

To process and transmit these signals, they are converted from an analog to discrete form. This conversion is likely achieved through an analog-to-digital converter (ADC) or a similar mechanism. The discrete signals, which are essentially digital representations of the original analog signals, can then be easily manipulated and transferred.

The discrete signals are passed through a discrete signals interface, which could be a set of electrical connections or components designed to facilitate the transfer of digital signals. This interface allows the discrete signals to be communicated to other systems or devices for further processing or analysis.

By employing this process of converting analog signals from the fly's eye into discrete signals, it becomes easier to handle and manipulate the information obtained from the optical sensors, enabling subsequent stages of signal processing, analysis, or integration with other systems.

In real life, two fly eyes simultaneously cover the entire sky, eliminating the need for wide-angle scanning. They can send signals about threat positions practically instantaneously, without requiring any processing. The position of the sensor already contains information about direction.

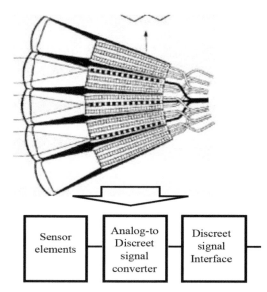

Sensor elements	Analog-to Discreet signal converter	Discreet signal Interface

Figure 1.1 A picture and structure of a fly's eye. Multiple angularly spaced optical sensors produce analog signals proportional to optical signals. These analog signals are then converted into discrete signals, which are transferred through a discrete signal interface.

The fly's eye architecture can be utilized for advanced antenna array designs. An antenna array (see Figure 1.2) comprises multiple angularly tilted directional antennas, which can be positioned closely or distributed along the perimeter of a ground vehicle or aircraft. Each antenna (sensor) is connected

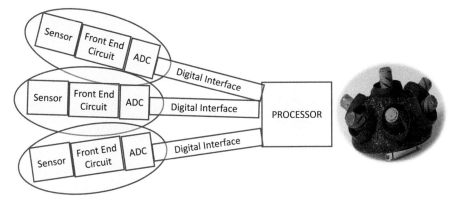

Figure 1.2 Nature-inspired fly eye antenna array, which must comprise multiple angularly tilted antenna sensors. Each sensor is coupled with a front-end circuit that includes a detector and a gain-controllable amplifier. After digitizing the signals from each antenna, they are transferred to the processor (brain) through a digital interface.

to a front-end circuit, which typically comprises a detector and a gain-controllable amplifier. The detector is responsible for converting the received electromagnetic signals into electrical signals that can be further processed. The gain-controllable amplifier allows for adjusting the signal amplification, which can help optimize the signal-to-noise ratio or adapt to different environmental conditions. Each antenna is coupled with a receiving/transmitting front-end circuit and connected to the processor by a digital interface.

Once the signals from each antenna are processed by the front-end circuitry, they are digitized. Digitizing the received signals directly on each directional antenna, relative to a common synchronization source (processor sampling signals), allows for the distribution of antennas and the recreation of a wave front or real-time digital hologram in the processor.

After digitization, the signals from each antenna are transferred to the processor, often referred to as the "brain." This transfer is accomplished through a digital interface, which could be a set of communication lines, protocols, or interfaces designed to facilitate the exchange of digital data between the antenna array and the processor.

Transformation and processing of received signals in the time domain, frequency domain, and multi-axis space domain will provide additional possibilities to enhance communication and visibility quality and reliability. The signals from each antenna are digitized and then transferred to a processor through a digital interface, enabling further processing and analysis of the maximum possible received data.

An antenna array with multi-beam antennas enables continuous wide coverage, allowing for the reception of maximum non-interrupting information from all signal sources or targets simultaneously. The multi-axis 3D coverage provided by the array facilitates target recognition and ensures reliable reception with high data rates for multiple links in communication systems.

The use of fast one-step algorithms in simultaneous multi-channel signal processing enables high-speed detection, tracking, and communication. Additionally, the application of signals from reference antennas can help suppress clutter and noise.

By employing a multi-beam antenna array with angularly tilted directional antennas, it is possible to achieve full sky coverage without the need for mechanical or electronic rotation or scanning [1, 2].

1.2 Coverage Area

The coverage area is an important parameter for an antenna array. In the case of a fly, its two eyes with angularly tilted sensors cover the entire sky, providing safety and situational awareness. An antenna array that covers the entire sky can be used for multi-orbit reliable communications or for drone/missile protection. Let us compare the coverage areas of mechanical rotating antennas, electronically scanning antennas, and fly eye antenna arrays (see Figure 1.3).

Mechanical rotating antennas are commonly used to provide a 360° coverage in azimuth, but they have limitations in terms of elevation coverage. They have blind zones in the upper hemisphere where targets or signals cannot be detected.

Active electronically scanned arrays (AESAs) offer good target detection capabilities in the front area, but they also have limitations. The angle of arrival (AOA) for AESAs is limited both in azimuth and elevation, which means they may not cover the entire sky or a full sphere without gaps or blind zones.

In contrast, a fly eye antenna array, as described earlier, can indeed provide coverage of the entire sky or a full sphere without any blind zones. This type of antenna array comprises multiple angularly tilted antennas or sensors, resembling the structure of a fly's eye. By combining the signals received from these antennas, a fly eye antenna array can offer practically instantaneous information about multiple targets or signal sources.

Fly eye antenna arrays have advantages in terms of their ability to provide a 360° coverage in both azimuth and elevation, making them suitable for applications such as radars, communication systems, or navigation systems where comprehensive situational awareness is crucial.

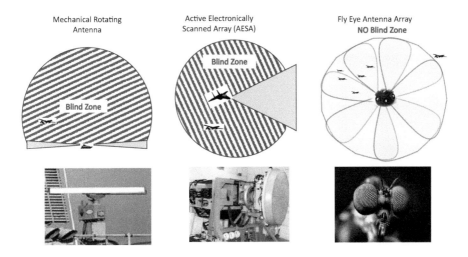

Figure 1.3 A comparison of blind zones (areas where the antenna array cannot "see" a target) for a mechanical rotating antenna, an active electronically scanned array (AESA), and a fly eye antenna array.

1.3 Target Illumination Time: Thread Detection

The detection time for threats is crucial for the survival of a fly, and the fly eye architecture provides minimal time for threat detection and direction finding. In radar systems, the target detection time depends on the duration of target illumination. This detection time includes the time it takes to scan the area of observation and the time of target illumination. Longer target illumination time leads to more reliable target detection but decreases the accuracy of direction finding for scanning beam systems. In communication systems, the term "active connection time" is often used, which is proportional to the link budget.

A mechanical rotating antenna requires a full 360° scan to detect all targets in azimuth. This process usually takes between 0 and 20 seconds or around 1 second for the fastest radars (see Figure 1.4).

In AESA, a narrow beam provides better directional accuracy for target detection. A smaller beam spot yields better direction to the targets. However, the narrow beam needs to be scanned across the entire area of observation to detect all targets. This scanning process typically takes approximately 0.5–1 second to scan an area of around 120° × 60°.

In a fly eye radar, the multi-beam antenna array covers the entire sky, with each antenna coupled to a separate channel connected to the processor.

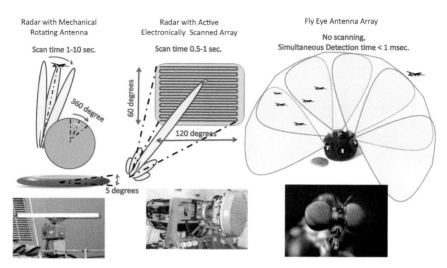

Figure 1.4   Comparison of time of detection target for mechanical rotating radar, AESA, and fly eye radar.

This means that the fly eye antenna array can provide information about all targets simultaneously and continuously, without interruption. The detection time for the fly eye radar consists of the time for detection, which is approximately a few microseconds for a semiconductor detector, and the time for direction finding through signal processing. Typically, a one-step algorithm is used to calculate amplitudes or phase ratios in the two channels, which can take a few milliseconds depending on the processor or FPGA (Field Programmable Gate Array) being used. There may be exceptions if multiple targets are located within the area of observation of a single antenna, which could require more complex processing algorithms.

1.4 Target Illuminating Time: Line Budget

The target illuminating time is an important parameter for radars as it allows for the comparison of radar range, target position information, and information required for target recognition and identification. For communication systems, it represents the connection's line budget or the time required for a non-interrupted connection.

One complete scan of a mechanical rotating antenna only provides a single hit (short pulse reflected from the target) for each target. This means that a rotating antenna can provide one brief pulse of information (direction to the target and potential information for target recognition and identification) about each target every 1–10 seconds (see Figure 1.4).

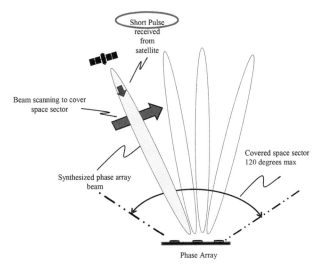

Figure 1.5 Illustration of how the scanning beam in a phased array can limit the satellite link budget. Switching the beam between multiple satellites can significantly worsen the link budget.

In the case of an AESA, it can provide one hit or short pulse of information for each target every 0.5–1 second (see Figure 1.4).

A radar with a mechanical rotating antenna can transmit and receive a maximum of one target hit pulse every 1–10 seconds, as one pulse hits the target per scan. It is evident that phased arrays with scanning beams limit the illuminating time (communication with satellites) and provide very short connection times, thus limiting the link budget (see Figure 1.5).

Antenna elements need to be switched for the observation of different space sectors and/or for multi-band functionality. This leads to a decrease in the time available for communication with a single satellite (link budget). Electronically steerable panel antennas, which are part of a phased array, also have limitations in the frequency band of the communication link. The steerable phase is directly related to frequency and a phased array cannot be inherently wideband or multi-frequency. The tradeoff between frequency band, gain, and active switching time leads to a loss in signal and/or data rate. Panel phased arrays require motors to change direction segments. On the other hand, an array of directional antennas allows for the solution of these problems. Direct digitizing and synchronization on each antenna enable the distribution of antennas on concave/convex surfaces, along the perimeter of a vehicle or among a swarm of manned/unmanned vehicles.

Figure 1.6 presents the fly eye staring array of directional antennas designed for simultaneous communication with LEO/MEO/GEO satellites.

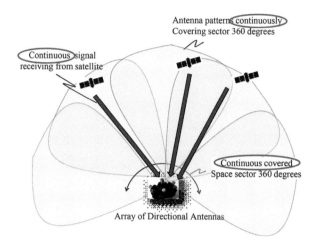

Figure 1.6   Illustration of an array of directional antennas capable of providing coverage across the required space. This array offers high gain in any direction, allowing for multiple beams and uninterrupted illumination of all targets or communication with multiple satellites while maintaining full link budgets.

The array consists of a multitude of directional antennas, enabling continuous communication with multiple satellites and maximizing the satellite link budget across a wide spatial area.

The fly eye antenna array architecture offers enhanced capabilities for receiving target information. For instance, if the distance to a target is 1 mile (which equals 5280 feet), the time for the reflected pulse to return is approximately less than 1 microsecond. Consequently, a pulse with a width of 1 microsecond can be transmitted and reflected from the target every 10 microseconds for each antenna. This means that the fly eye radar can transmit to and receive from any target direction at a rate of 100,000 pulses per second, significantly increasing radar sensitivity. The integration of these received 100,000 pulses will significantly boost radar range and provide detailed information about the target. The fly eye antenna array architecture also enables the enhancement of link budgets for communication systems across all of space.

1.5 Radar Range

Based on the radar equation, the maximum range of a monostatic radar can be estimated by considering various factors, including the radar target cross section (σ), wavelength (λ), transmitted-signal power (P_t), transmitting and receiving antenna power gains (G_t and G_r), and the pattern propagation factors (F_t and F_r) for the transmitting antenna-to-target path and target-to-receiving antenna path, respectively.

In the case of a wide beam width antenna array, it is true that the transmitted and received energy will be dispersed over a wider area. This can result in a decrease in the sensitivity and range of the radar compared to a narrower beam width antenna.

The maximum range equation for monostatic radar (one in which the transmitter and receiver are co-located) is given by the following equation [3]:

$$R = \left[\frac{P_t G_t G_r \sigma \lambda^2 F_t^2 F_r^2}{(4\pi)^3 P_r} \right]^{\frac{1}{4}} \tag{1.1}$$

where:

    R – radar-to-target distance (range);
    σ – radar target cross section;
    λ – wavelength;
    P_r – received signal power being equal to the receiver minimum detectable signal S_{min};
    P_t – transmitted signal power (at antenna terminals);
    G_t – transmitting antenna power gain;
    G_r – receiving antenna power gain;
    F_t – pattern propagation factor for transmitting antenna-to-target path;
    F_r – pattern propagation factor for target-to-receiving antenna path.

In the case of a fly eye radar with a wider antenna beam width, such as 10 times wider (increasing from 3° to 30°), there will be a spread of the transmitted energy over a larger area. With a wider beam width, the energy of the transmitted signal will be distributed over a 10 × 10 square instead of a smaller area. According to the inverse square law, the signal reaching the target will decrease by a factor of 100 due to the increased spreading of the energy over a larger area.

In terms of the maximum range, the fly eye radar with a wider beam width can transmit and receive pulses to and from any target direction at a much higher rate compared to a regular scanning antenna radar. The regular radar with a scanning antenna can typically transmit a maximum of 1 target hit pulse every 30–40 seconds, whereas the fly eye radar can transmit and receive approximately 100,000 pulses per second to and from any target direction. The integration of the received 100,000 pulses per second in the fly eye radar will indeed provide a significantly increased amount of information about the target.

It is important to note that the specific range performance of the fly eye radar with a wider beam width will depend on various factors, including the transmitted power, receiving sensitivity, antenna gain, propagation environment, and signal processing techniques utilized. Detailed analysis and simulations would be required to estimate the exact range and performance characteristics of the fly eye radar system under these conditions. The maximum range equation for fly eye radar must include the number of integrated pulses:

$$R = \left[\frac{(P_t \mathbf{I_e} \mathbf{M}) G_t G_r \sigma \lambda^2 F_t^2 F_r^2}{(4\pi)^3 P_r} \right]^{\frac{1}{4}} \qquad (1.2)$$

where:

I_e – integrator efficiency;

M – number of transmitted/received pulses per period of integration.

It appears that based on the equation and parameters provided above and according to equation (1.2) and the given conditions, if the number of transmitted pulses per second (M) is 100,000 and the transmitted power (P_t) is smaller by 100 times, the numerator in the equation will be larger by 1000 times. As a result, the maximum radar range will increase proportionally. This implies that the fly eye radar, with its ability to transmit smaller power in the target direction but continuously observe and integrate reflected signals, can achieve an increased radar range. By simultaneously correlating and integrating thousands of signals per second from each point of the surveillance area, the fly eye radar can not only detect low-level signals (such as low-profile targets) but also aid in signal recognition, classification, and target identification through the use of diversity signals, polarization modulation, and intelligent signal processing techniques.

It is important to note that the specific increase in radar range and the effectiveness of signal recognition and classification will depend on various factors, including the specific implementation of the fly eye radar system, the environment, and the signal processing algorithms employed. Extensive analysis, simulations, and experimental testing would be necessary to validate and optimize the performance of the fly eye radar under these conditions.

Figure 1.7 compares the range of detectable transmitted power (R_t) for a regular scanned radar with larger transmitting power and smaller beam width to the range of detectable power (R_t) for the fly eye radar with wider beam width and longer integration time. In the case of the regular scanned radar, with a larger transmitting power and smaller beam width, the range of detectable transmitted power (R_t) is expected to be larger. This is because the radar

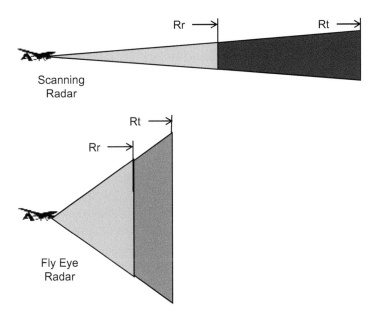

Figure 1.7 Range of detectable transmitted power R_t is larger for regular scanned radar, where larger transmitting power and smaller beam width. Range of detectable power R_r is approaching the range of detectable transmitted power because of the increasing time of integration.

system can focus more power into a narrower beam, resulting in a stronger received signal at longer ranges.

On the other hand, for the fly eye radar with a wider beam width and longer integration time, the range of detectable power (R_r) approaches the range of detectable transmitted power. This means that even with a lower transmitting power, the fly eye radar can achieve a comparable range of detectable power due to its ability to integrate signals over a longer period of time. The longer integration time allows the fly eye radar to accumulate and process multiple signals, improving the system's sensitivity and ability to detect weak signals.

The specific ranges of detectable transmitted power (R_t) and detectable power (R_r) will depend on various factors, including the specific parameters of the radar systems, the environment, and the signal processing techniques utilized. Further analysis and simulation would be necessary to determine the exact ranges and performance characteristics of the two radar systems under consideration.

The range of detectable target R_r will approach the range of detectable transmitted power R_t in the fly eye radar because a few orders larger number of target hit pulses and integration of signals (Figure 1.7). Therefore, radars with mechanical or electronically scanning narrow beam can provide larger detectable range but need more detection time for scanning the whole area of observation. The fly eye radar can provide smaller range, but wider area of observation and faster detection time, which is preferred, for example, for tactical systems for drone detection.

The range of detectable target R_r for a regular scanning radar with the same minimum receivable power of -100 dBm will depend on power reflected from the target or target cross section. The range of detectable target R_r will approach the range of detectable transmitted power R_t in the fly eye radar as a result of large number of signals and signals integration.

In the case of regular scanning radar, the range of detectable target (R_r) will depend on factors such as the power reflected from the target and the target's cross section. The range of detectable target (R_r) will approach the range of detectable transmitted power (R_t) in the fly eye radar due to the larger number of signals and the integration of those signals. The fly eye radar, with its ability to transmit and receive a larger number of pulses per unit of time, can provide a higher density of information and improve the detection capabilities.

Additionally, the smaller transmitted power for the fly eye antenna array can be compensated by installing additional transmitters for each observation section. This approach can help overcome the decrease in transmitted power by distributing the transmitters throughout the array, effectively increasing the overall power transmitted by the system. Furthermore, the increased range of detectable target in the fly eye radar can be beneficial in certain applications, such as the concept of an invisible radar. By detecting the high-power scanning beam of an intruder radar at an earlier stage, the fly eye radar can help avoid approaching the detectable range of the intruder radar, providing a tactical advantage (Figure 1.8).

1.6 Passive Radar Concept

It appears that radar visibility can be decreased by reducing transmitted power and increasing integration time for non-interruptive signal reception (Figure 1.9). In this context, a passive radar system can utilize existing signals such as those from satellites, broadcasts, communications, or cell phones. These signals are typically of low power, and the receiving system would need to be enhanced to reliably detect small cross-sectional targets. To increase the information available for target detection or recognition, there are several methods that can be employed. One approach is to receive

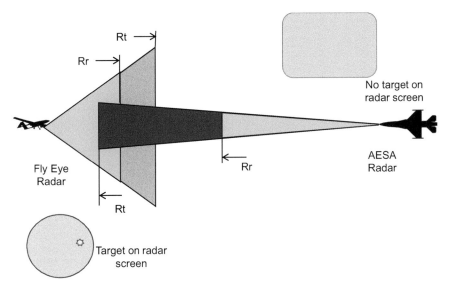

Figure 1.8 Invisible radar concept.

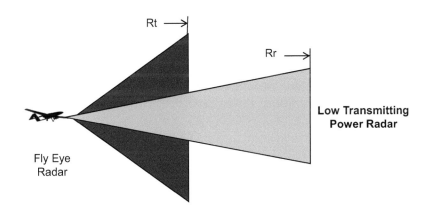

Figure 1.9 The range of detectable transmitted power R_t will be smaller than the range of detectable target R_r for a low-power radar if correlation of received signals is applied in addition to integration. Low transmitting power radar can decrease UAV visibility for enemy radars.

signals from a multi-axis arranged antenna array, which can improve the spatial coverage and capture signals from various directions. Additionally, using multiple frequencies simultaneously, sequentially, or randomly can provide diversity and increase detection reliability.

In terms of waveform and modulation, employing smart waveforms that are optimized for specific detection scenarios can enhance the radar system's

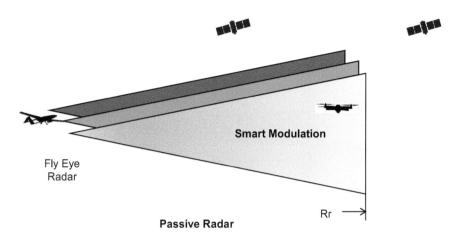

Figure 1.10   Smart modulation (compression of signals, modulation of signals polarization, and multi-frequency processing) will allow to create a passive radar concept.

performance. Domain transformations of received signals, such as Fourier or wavelet analysis, can also be utilized to extract relevant features and improve detection and recognition capabilities.

These techniques aim to increase the sensitivity and reliability of the radar system in detecting and recognizing targets while minimizing its own visibility and reducing the transmitted power.

Passive, not transmitting, radars can use satellites, broadcast, different communications, or cell phones transmitting signals. These signals are of low power and the receiving system needs to be enhanced for reliable detection of small cross-sectional targets. There are a few methods for increasing the information about the target for detection or recognition. Signals can be received from multi-axis arranged antenna array, by using a few different frequencies simultaneously, sequentially, or randomly. Some kind of smart waveform, modulation, and domain transformation of received signals can increase detection reliability (Figure 1.10).

1.7 Phase Array vs. Fly Eye Antenna Array

The challenge of future mobile antenna array technology for all-space applications is to develop a versatile solution that can meet the requirements of various orbits such as MEO, LEO, and VLEO (middle Earth orbit, low Earth orbit, and very low Earth orbit) and support different applications like communication, navigation, counter UAS, or radar systems. Here are the formalized requirements for such antenna arrays:

- The antenna array must simultaneously cover the entire sky to "see" all possible signals/threat sources.

- For maximum efficiency, sensitivity, and range, the antenna array must provide continuous staring (no scanning and no switching) for each signal source and provide non-interrupting line's budget.

- For full performance lines or fast threat detection and tracking, the antenna array must provide simultaneous processing of signals from all signal sources, be it multi-channel, optional multi-frequency, or multi-function.

- The antenna array must be solid-state (no moving parts) and provide jam, spoof, and EMP protection.

Isotropic antenna elements in planar phase array are limiting the covered space sector. A phase antenna array with one scanning beam required minimum half of a second for scanning all covering sector. Time of communication with one satellite will be very small. A larger number of satellites can be detected and tracked with the multi-beam antenna system. But the number of beams decreases the gain of each separate beam for one antenna array with a fixed aperture size. For phase array, the gain of antenna array with some fixed apertures in one direction equals G. For two directions/beams, aperture is divided into two folds for each beam, and gain decreases by half, leading to decreasing antenna array sensitivity and range.

The main advantage of the fly eye antenna array is that it can provide a full coverage of the entire sky or a required area of observation without the need for beam scanning or switching. This is particularly beneficial for applications that require coverage of multiple orbiting satellites or the detection of swarms of individual UAVs (Figure 1.11).

By eliminating the need for scanning and providing continuous coverage, the fly eye antenna array offers enhanced performance, faster threat detection and tracking, and improved sensitivity and range compared to traditional scanning or switched-beam systems.

One of the advantages of fly eye antenna arrays is their flexibility in the arrangement of directional antennas. Unlike traditional antenna arrays that require a half-wavelength spacing between antenna elements, the directional antennas in fly eye arrays can be arranged in small-sized modules or loosely distributed around the perimeter of a UAV or vehicle. This flexibility is particularly valuable for long-wavelength RF range systems. The arrangement of the antennas allows for overlapping antenna patterns in different directions, providing enhanced coverage and flexibility. This can be seen in Figure 1.11, where the antenna patterns overlap in one direction, and Figure 1.12, where

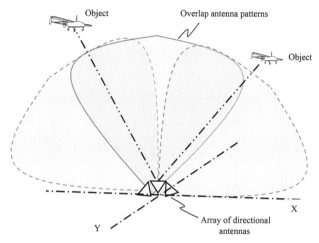

Figure 1.11 Ground-based all-space antenna array with *X*-axis overlapping antenna patterns.

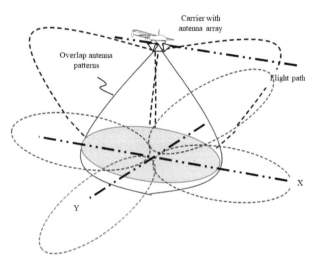

Figure 1.12 Airborne carrier-based all-space antenna array with two-directional *X*- and *Y*-axis overlapping antenna patterns.

they overlap in two directions. In some cases, fly eye arrays can even achieve overlapping antenna patterns in multiple directions, providing multi-axis coverage.

The compact size of the individual directional antennas and the ability to position them on airborne, sea, or ground vehicles make fly eye arrays suitable for various applications. The additional power required for controlling the beamformer and beam steering is manageable, considering the benefits of

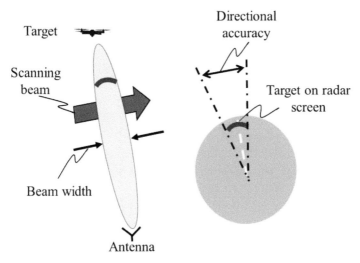

Figure 1.13 Directional accuracy of scanning antenna beams limited by beam width for mechanical scanning antennas or phase array electronically scanning antenna beams.

the continuous coverage and non-interrupting communication or surveillance capabilities provided by fly eye antenna arrays.

The arrangement and distribution flexibility as well as the small size and controllable beamforming capability of fly eye antenna arrays make them well-suited for use in different vehicles and scenarios where space is limited and efficient coverage is required.

In optical systems, the diffraction-limited angular resolution is determined by the wavelength of the light and the diameter of the objective's entrance aperture [4]. The diffraction limit sets a fundamental limit on the angular resolution of the system.

For radar systems, the minimal beam width is also limited by the beam spot size, which is determined by the wavelength of the radar signal. Higher frequency radar signals, such as those in the millimeter-wave range, allow for smaller beam widths and increased beam gain. This is achieved by employing a larger number of antenna elements in the array, ranging from hundreds to thousands. However, as the frequency decreases, the beam width becomes larger, and the directional accuracy of the antenna system decreases and the antenna array becomes larger because larger spacing is required between the antenna elements at lower frequencies. As a result, directional accuracy drops and the antenna system becomes physically larger in order to accommodate the larger spacing between elements, as shown in Figure 1.13. It is important to note that high directional accuracy is crucial for both optical and radar systems. In the case of radar phase arrays, even a small error in the direction of a narrow beam can lead to a significant drop in signal gain and a reduction in array sensitivity and range. The diffraction limit and the tradeoff between frequency, beam width, and directional accuracy play important roles in determining the performance and physical size of optical and radar antenna systems.

Directional accuracy in antenna systems is often measured by analyzing the phase delay between two antennas positioned with a certain baseline distance in a plane perpendicular to the direction of the target, as shown in Figure 1.14(a). The accuracy of the phase measurement is limited by the baseline distance between the antennas.

To improve the directional accuracy, monopulse techniques are commonly used. In a monopulse array, the antenna patterns of multiple elements overlap, as shown in Figure 1.14(b). This overlapping allows for more accurate measurement of the direction of arrival by calculating the ratio of amplitudes or phases from different elements.

The advantage of the monopulse method is that it does not require a specific baseline distance between antennas. As a result, the antennas can be arranged in a smaller space, which is beneficial when size constraints are a concern. Furthermore, in the monopulse method, each antenna is coupled with separate signal processing channels, enabling faster detection and reducing processing time. This is particularly important for applications such as hypersonic missile and UAV detection, where rapid and accurate detection is crucial. By employing monopulse techniques and utilizing separate signal processing channels, antenna systems can achieve improved directional accuracy and faster detection capabilities.

Antenna arrays with multi-beam and multi-axis overlapping patterns offer significant advantages in terms of receiving maximum information from all signal sources or targets simultaneously and maximizing the line budget. By providing continuous wide coverage and staring capabilities, these antenna arrays enable the reception of signals from multiple directions without the need for scanning.

The multi-axis 3D coverage of antenna arrays enhances the reception of signals and provides a greater amount of information from signal sources. By overlapping antenna patterns in multiple directions, the array can receive signals from various angles, increasing the overall sensitivity and data rate.

The ability to receive maximum information from all signal sources simultaneously is particularly valuable in scenarios where multiple targets or signal sources need to be monitored and tracked in real time. This can be beneficial for applications such as communication systems, radar systems,

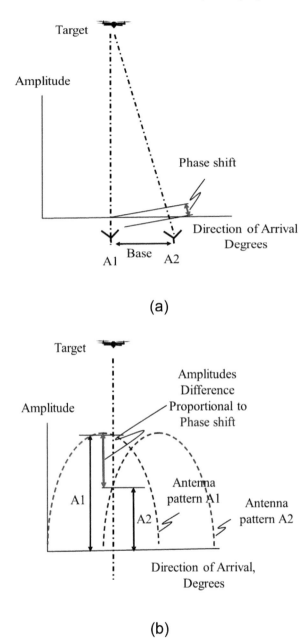

Figure 1.14   (a) Regular direction of arrival finding by measurement of phase shift in two points with base distance between them. (b) A more accurate direction of arrival measurement can be done by calculating the ratio of amplitudes in two antennas with overlapping antenna patterns by using the monopulse method.

or surveillance systems where a comprehensive view of the surrounding environment is essential. The continuous wide coverage and high data rate capabilities of these antenna arrays enable reliable reception and efficient communication. The ability to receive signals from multiple directions simultaneously ensures robust and uninterrupted communication links.

Overall, antenna arrays with multi-beam and multi-axis overlapping patterns and staring capabilities provide significant advantages in terms of information reception, line budget optimization, and reliable high-speed communication.

A review of the benefits of the fly eye antenna array and the most advanced digital beam forming (DBF) antenna array [5] is shown in Table 1.1.

1.8 Conclusion

In summary, a brief comparison between the nature-inspired fly eye antenna array and regular beam scanning antennas and antenna arrays leads to the following conclusions:

- Fly eye antenna arrays do not require digitally applied time delays, as each antenna already covers a specific space sector. This eliminates the need for high power consumption and large-sized beamformers.

- The gain and sensitivity of regular beam scanning arrays decrease proportionally with the increasing number of simultaneous scanning beams. In contrast, the gain of each antenna and coupled channel in a fly eye array is constant or can be automatically controlled, irrespective of the number of active channels.

- Electronically scanning beams still require time to cover the scanning sector, while fly eye antenna arrays provide instant detection of any target or connection channel. This significantly reduces detection time, making them suitable for applications such as hypersonic missile and UAV detection.

- Staring non-scanning antennas in fly eye arrays dramatically increase the illumination time for targets in radars or improve the link budget for communication systems, as there is no need for channel switching. Increasing the number of targets (channels) proportionally decreases the illuminating time in regular beam scanning systems.

- Adaptive antenna array processing typically requires a significant amount of time and is suitable for low-power signals to prevent receiver

Table 1.1 Benefits of Fly Eye antenna array vs Phase array with digital beamforming.

Phase array with digital beamforming	Fly eye antenna array
Flexibility	Flexibility Automatic gain control for
Digitally applied time delay	each beam
	No time delay control block
Multiple simultaneous beams	Multiple beams simultaneously cover the
Decreases gain/sensitivity for constant	entire sky
aperture	Constant or controllable gain in each
	beam
Faster search	Monopulse high-accuracy direction
Electronically scanning beams	finding
	No scanning, no beams interrupting
Tracking multiple targets	Tracking multiple targets
Dramatically decreases target	Non-interrupting continuous illuminating
illuminating time by scanning beam	of (tracking) multiple targets
Adaptive antenna array processing	Monopulse one iteration processing of
Adaptive processing is possible if the	signals in overlapping beams as a ratio
first signal does not saturate or destroy	of phases or amplitudes
all antenna elements	Distribution of directional antennas
Jammer suppression	around vehicle perimeter created
Thousands of iteration suppressions	additional jam protection of the system
increase the target detection time	
Noise improvements	Noise improvement by the application of
	reference signals

saturation. On the other hand, fly eye antenna arrays require minimal calculation time for direction finding and provide additional jamming protection from high-power jam signals due to their directional antennas.

- The use of reference signals in overlapping or reference antennas in fly eye antenna arrays improves and speeds up noise suppression in a wide dynamic range.

The fly eye antenna array offers advantages such as instantaneous detection, continuous coverage, constant gain, fast processing time, and enhanced jamming protection, making it a promising technology for various applications in communication, radar, visualization, and surveillance systems.

References

[1] P. A. Molchanov, A. Gorwara, "Fly Eye radar concept". IRS2017. International Radar Symposium, Prague, July 2017.

[2] P. A. Molchanov and O. V. Asmolova. *All-digital radar architecture*. Conference: SPIE Security + Defense, DOI: 10.1117/12.2060249, October 2014.

[3] Radar Handbook. M. I. Skolnik. Chapter 1. p.62, McGraw Hill, 1990.

[4] Diffraction-limited system limit. Wikipedia.

[5] Jon Bentley and Jerome Patoux, Analog Devices, Wilmington, Mass." The Continuing Evolution of Radar, From Rotating Dish to Digital Beamforming", Microwave Journal, eBook, p. 5–8, January, 2023.

2

Evolution of Antenna Array Systems

2.1 Omnidirectional Antennas

Omnidirectional antennas are designed to radiate or receive radio signals equally in all directions around the antenna's axis, in the azimuthal plane. However, their radiation pattern in the elevation plane is not uniform and typically forms a donut-shaped pattern. This means that there are blind zones or areas of reduced sensitivity above and below the antenna.

The donut-shaped radiation pattern of an omnidirectional antenna means that it provides better coverage horizontally, while the coverage vertically is limited. This characteristic is essential to consider when deploying and positioning omnidirectional antennas in order to ensure optimal signal reception or transmission.

In certain applications, such as wireless communication networks or broadcast systems, the donut-shaped radiation pattern of omnidirectional antennas is desirable as it allows for broad coverage in the horizontal plane. However, in situations where specific directional coverage is required or there is a need to minimize the risk of interception, the use of directional antennas becomes crucial.

Directional antennas, such as Yagi antennas or parabolic reflector antennas, have focused radiation patterns that can be adjusted to provide a narrower beamwidth and more precise coverage in a specific direction. This allows for increased gain and improved reception or transmission in the desired direction while reducing sensitivity in other directions. This characteristic of directional antennas can help minimize the risk of interception and enhance the security of the transmitted signals.

The first omnidirectional (isotropic, same gain in all directions) antenna was operated with a simple long piece of wire. In the early days of radio design, when low-frequency and long-wavelength signals were used, simple long wires were used as antennas. The length of the wire was determined by the wavelength of the signal, and it was often a half-wavelength long. At that time, people believed that such antennas would transmit and receive

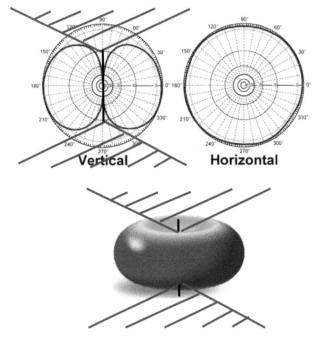

Figure 2.1   Half-wavelength omnidirectional antenna arranged as a piece of wire and its antenna patterns in vertical and horizontal planes.

signals equally in all directions – hence, the term "omnidirectional." It is called "long" because people used low-frequency long-wavelength signals on the beginning of first radio design. The wire became shorter with a shorter wavelength, and half of the wavelength was added later, when people determined how long the antenna must be and designed short wave radios. People normally use a half-wavelength-long piece of wire as an omnidirectional antenna. Why omnidirectional? Because people think the antenna transmits and receives signals omnidirectionally. However, it is important to note that the assumption of perfect omnidirectionality is not always accurate. While an antenna can radiate signals in a 360° pattern in the azimuthal plane, the radiation pattern in the elevation plane is not uniform. Instead, it forms a donut-shaped pattern, as shown in Figure 2.1. This means that there are blind zones or areas of reduced sensitivity above and below the antenna.

The presence of blind zones becomes more critical when the received signals are weak or close to the lower detection threshold. In such cases, the antenna's limited angle of view in the elevation plane can lead to reduced sensitivity and the potential for missed or weak signals. This is particularly important in applications where the detection of low-level signals or low radar cross-sectional targets is necessary. Antennas cover a 360° area by azimuth but provide a limited angle of view (AOV) by elevation with huge blind zones on top and bottom. Wide blind zones are not so important until transmitting and/or receiving signals higher than the lower receiving signal threshold. But blind zones must be taken into account if signals are approaching the lower threshold. It is important for receiving low-level signals detecting low radar cross-sectional targets. In radio communication, the term "omnidirectional antenna" refers to a class of antennas that radiate equal radio power in all directions perpendicular to an axis (azimuthal directions). However, the power varies with the angle to the axis (elevation angle) and declines to zero on the axis itself [1, 2]. Understanding the limitations of omnidirectional antennas and considering the specific requirements of an application are crucial for achieving optimal performance.

It is important for radio and antenna designers to consider the limitations of omnidirectional antennas and the specific requirements of their application. In scenarios where the elevation angle is crucial for signal reception, alternative antenna designs or additional antenna systems may be needed to ensure reliable detection and reception of signals from all directions.

2.2 Directional Antenna Arrays

Directional antennas, also known as anisotropic antennas, are designed to have different gains in different directions. They are often implemented as antenna arrays, consisting of multiple antenna elements or multiple antennas working together to achieve a desired radiation pattern or beam. Unlike omnidirectional antennas that radiate equally in all directions, directional antennas concentrate their radiation pattern or beam in a specific direction or a few directions. This results in a narrower angle of view (AOV) in both azimuth and elevation, as shown in Figure 2.2.

The wide blind zone associated with directional antennas is not necessarily a disadvantage. In fact, it can be beneficial for certain applications. For example, in target detection scenarios, a smaller beam spot corresponding to better directional accuracy is desired. By narrowing the beam, the antenna can focus its sensitivity and gain on specific areas, increasing the likelihood of detecting targets or objects within that region.

Additionally, directional antennas are often used for transmitting narrow beams with maximum gain to achieve long-distance communication or radar applications. By concentrating the transmitted power into a narrow beam, the antenna can improve the signal strength and coverage in the

DIRECTIONAL ANTENNA

Figure 2.2 Sample of log-periodic, multi-element directional antenna designed to operate over a wide band of frequencies.

desired direction, enabling long-range detection or efficient communication with specific targets.

It is important to note that the choice between omnidirectional and directional antennas depends on the specific application requirements. While omnidirectional antennas provide coverage in all directions, directional antennas offer higher gain, improved range, and better directional accuracy, making them suitable for tasks that demand targeted and focused radiation patterns.

2.3 Phase Arrays

The concept of a phased array, which allows for steerable transmission and reception of radio waves, was first demonstrated by Karl Ferdinand Braun, a Nobel laureate, in 1905. He showed that by properly phasing multiple antenna elements, the transmitted radio waves could be enhanced in a specific direction, resulting in improved signal strength and range in that direction. During World War II, steerable phased array transmission systems were used to assist in the landing of aircrafts. By adjusting the phase of the signals in the array,

Figure 2.3   Karl Ferdinand and his phase array.

the transmission beam could be directed toward the aircraft, providing guidance and navigation assistance.

Phased array technology has since found various applications, including radio astronomy and radar systems. In radio astronomy, phased arrays are used to create interferometric radio antennas, where signals from multiple antennas are combined to achieve high-resolution imaging and directional sensitivity, as shown in Figure 2.3.

In radar systems, phased arrays offer the ability to electronically steer the radar beam without physically moving the antenna. This allows for rapid scanning, tracking of multiple targets, and adaptive beamforming for improved detection and discrimination capabilities. The development and generalization of phased array technology have significantly impacted various fields, enabling advanced communication systems, precise radar imaging, and scientific research in radio astronomy [3].

A phased array antenna consists of an array of antenna elements, each powered by a transmitter. The phases of the signals transmitted by each

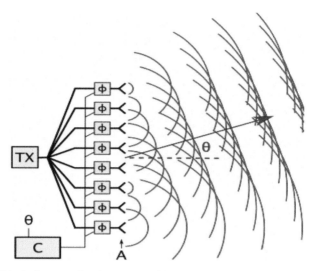

Figure 2.4 Block diagram of a phase array with computer controllable phases of each transmitted signal.

antenna element are controlled individually by a computer. By adjusting the phases of the transmitted signals, the spherical wave fronts emitted by each element can be superposed and combined to create a plane wave that travels in a specific direction. Phase array consists of an array of antenna elements A powered by a transmitter TX. Phases of each transmitted element φ are controlled by a computer C (Figure 2.4). The process of superposing coherent waves with the same frequency is called interference. In the case of a phased array, the phase shifters in the antenna system introduce a progressive delay in the radio waves as they move up the line of antenna elements. This delay causes each antenna to emit its wave front slightly later than the one below it, resulting in constructive interference in a specific direction.

By controlling the phase shifts of the transmitted signals using a computer, the angle θ of the beam can be instantly changed. This allows for the electronic steering of the beam without physically moving the antenna. The ability to rapidly change the beam direction makes phased arrays highly versatile and suitable for applications such as radar systems, where the beam needs to be quickly scanned or directed toward different targets. Phased arrays offer the advantage of beam agility and flexibility, allowing for adaptive beamforming, beam scanning, and beam shaping, all controlled by adjusting the phases of the transmitted signals.

Phased arrays can be configured with multiple directional antenna elements or antennas. In such a configuration, the phased array consists of a

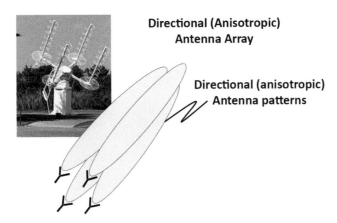

Figure 2.5 Antenna array comprising four directional (anisotropic) antennas focusing beams to some space point.

set of directional antennas, each with its own phase adjustment capability. This arrangement allows for separate adjustment of the antenna patterns for compensation of errors and temperature dependence in the transmitting lines.

By using directional antennas in a phased array, higher gain and longer range/sensitivity can be achieved compared to phased arrays with omnidirectional antenna elements. This is because directional antennas can concentrate the transmitted energy into a narrower beam, thereby increasing the effective radiated power in the desired direction. Phased array needs to compensate for the reduced gain or coverage due to the use of omnidirectional antenna elements; it can be achieved by increasing the number of antenna elements in the phase array. By adding more antenna elements, the overall gain and coverage of the phased array can be improved. The specific configuration of the phased array, including the number and arrangement of the directional antennas, can be tailored to the desired application and performance requirements (Figure 2.5).

Active electronically scanned arrays (AESAs) are computer-controlled antenna arrays that allow the beam of radio waves to be electronically steered in different directions without physically moving the antenna. AESA technology enables fast scanning of narrow beams, typically within a time frame of approximately 0.5–1 seconds. The scanning capability is limited by the azimuth and elevation angles of the plane antenna arrays.

AESA systems are often configured as arrays of omnidirectional antenna elements integrated into a planar panel. This arrangement helps to decrease the overall size of the array while still providing a wide coverage area for observing forward aircrafts. The large number of omnidirectional

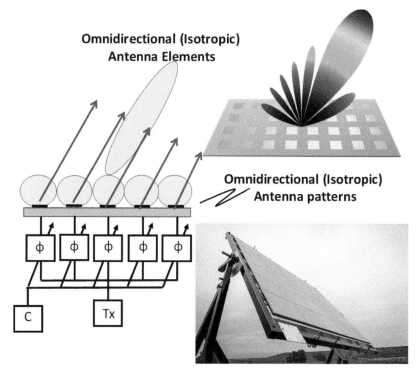

Figure 2.6   Phase array comprising an array of omnidirectional antenna elements connected with a transceiver and a computer controllable beamformer.

antenna elements in an AESA allows for the generation of narrow, high-gain beams for target detection.

One advantage of the AESA technology is that it can be implemented in a relatively small space volume for airborne applications, even with a large number of antenna elements ranging from a few to thousands. This is made possible by using small millimeter-wave signals, which have shorter wavelengths. It is worth noting that longer wavelengths are often used in radar systems that require better penetration of clouds or foliage. However, the size of the phase antenna array increases dramatically with a half-wavelength distance between antenna elements. Therefore, for longer wavelengths, the physical size of the antenna array may become a limiting factor. The AESA technology provides the capability for electronically scanning narrow beams with high-gain performance, making it suitable for various applications such as radar systems on aircrafts (Figure 2.6).

The gain of a phase antenna array can be affected by the number of antenna elements used. As the number of elements increases, the overall gain

Table 2.1 Beam steering gain on satellite signals.

Number of elements	Gain (dB)	Gain of one element (dB)
7	8.5	1.21
16	12	0.75
109	20.4	0.187

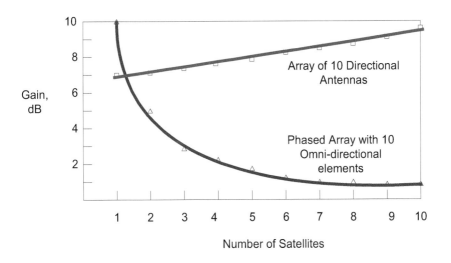

Figure 2.7 Gain dependence from a number of GPS satellites for a directional antenna array with 10 directional antennas and a phase antenna array with 10 omnidirectional elements.

of the antenna array increases. However, the gain of each individual element (efficiency) decreases with the increasing number of elements due to phase summing errors. This means that the gain achieved by each element becomes smaller as more elements are added to the array.

Table 2.1 provides an example of the gain values for different numbers of antenna elements in a phase antenna array. As shown, for a 7-element array, the gain of one element is 1.21, while for a 109-element array, the gain of one element decreases to 0.187 [4]. The size of the phase antenna array also increases proportionally with the number of antenna elements. In a phase array, the antenna elements need to be positioned at a distance of half wavelength, and each element requires a precise phase control. Small phase control errors in a large number of elements can significantly decrease the gain and coverage area of the antenna array.

Figure 2.7 illustrates the dependence of gain on the number of GPS satellites for two different types of antenna arrays: a directional antenna array with 10 directional antennas and a phase antenna array with 10 omnidirectional elements. It shows how the gain varies with the number of satellites, highlighting the potential tradeoffs between the gain and the number of antenna elements in different array configurations.

Phase arrays are typically designed for a specific frequency or a narrow frequency band. This is because the distance between antenna elements in a phase array is determined by the wavelength of the operating frequency. If the frequency changes, the wavelength changes, and the required phase delay between elements will be affected, leading to phase errors and degradation of the beamforming performance.

The frequency bandwidth of a phase array can be extended, but it often comes at the cost of tradeoffs between gain, directivity, and beamforming performance. Wider frequency bandwidths can result in a decrease in gain and directivity, which means that the antenna's ability to focus the signal on a specific direction is reduced. To maintain beamforming accuracy over a wider frequency range, more complex and sophisticated beamforming techniques and algorithms may be required, increasing the size and power consumption of the antenna system.

Furthermore, when it comes to planar phase arrays, there can be limitations in terms of coverage area near the edges of the array. This is because the phase control errors tend to increase toward the edges, which can affect the beamforming accuracy in those regions. Careful design considerations and calibration techniques are necessary to mitigate these issues and ensure uniform performance across the entire coverage area of the antenna. The frequency bandwidth and edge effects are important considerations in the design and operation of phase arrays, and tradeoffs must be made between gain, directivity, frequency range, and system size to meet the specific requirements of the application (Figure 2.8).

The coverage area of a planar phase antenna array is limited by the beamwidth of one antenna element due to the phase steering errors that occur toward the edges of the array. As the beam is steered away from the broadside direction, the phase errors become more pronounced, leading to reduced beamforming accuracy and coverage.

This limitation is illustrated in Figure 2.9, where the coverage area of the planar antenna array is shown to be smaller than the beamwidth of a single omnidirectional element. The influence of phase shift errors toward the edge of the array restricts the effective coverage area.

As a result, planar phase antenna arrays may not be suitable for applications that require horizon-to-horizon (180°) tracking or wide-angle coverage. The tilt of the planar array can help increase coverage in one direction

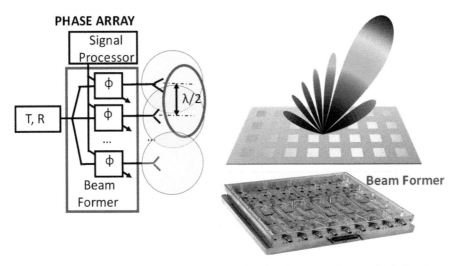

Figure 2.8 Phase array with an array of omnidirectional antenna elements including large size and consumption power beamformer block. Half-wavelength distance between antenna elements increases phase array size.

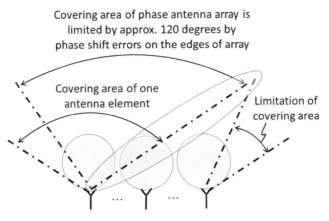

Figure 2.9 Covering area of a planar antenna array smaller than the beamwidth of one omnidirectional element because of the influence of phase shift errors on the edge of the antenna array.

but may exclude coverage from other directions, further limiting the overall coverage area. It is important to consider these limitations when designing and deploying planar phase antenna arrays, and alternative antenna configurations or techniques may be needed to achieve a wide-angle coverage or horizon-to-horizon tracking (Figure 2.9).

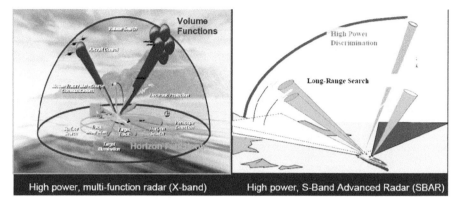

Figure 2.10   Proposed by ONR multi-function radar diagrams, applied for detection of multiple targets in different space sectors.

2.4 Multi-beam Phase Array

The detection and tracking of multiple targets with different altitudes, space sectors, and speeds, as well as the increasing number of drones and various space communication systems, pose significant challenges. To address these challenges, multi-beam antenna arrays can be used.

Multi-beam antenna arrays have the capability to generate and steer multiple beams simultaneously, allowing for the tracking of multiple targets or the coverage of different ranges and sectors of space. This enables the array to provide continuous staring, meaning that it can maintain coverage on multiple targets without the need for scanning or switching between them.

By utilizing multiple beams, the antenna array can track and communicate with multiple satellites simultaneously. Each beam can be directed toward a specific satellite or target, allowing for simultaneous processing of signals from multiple sources. This multi-channel, multi-frequency capability is essential for efficiently managing the communication and tracking requirements of various space systems.

With a multi-beam antenna array, it becomes possible to have a single antenna system and terminal that can cover the entire sky and provide continuous staring while simultaneously processing signals from multiple satellites or targets. This helps to overcome some of the challenges associated with the detection and tracking of multiple targets and the diverse requirements of space communication systems (Figures 2.10 and 2.11).

But multi-beam antenna systems still face certain challenges. Here are some additional considerations:

1.   Scanning time: While multi-beam systems can provide coverage over a wider area compared to single-beam systems, scanning the entire

Figure 2.11 Diagram of a multi-beam system for the detection of a swarm of drones proposed by SRC Inc.

semi-sphere area can still take time. The speed of scanning depends on factors such as the number of beams, beamwidth, and the desired angular resolution. Design tradeoffs need to be made to strike a balance between scanning speed and directional accuracy.

2. Beam interference: Increasing the number of beams in a multi-beam system can lead to interference between beams. Each beam requires its own resources, including power and bandwidth. Allocating resources to multiple beams can decrease the available resources for each beam, potentially impacting their individual performance.

3. Vulnerability to jamming: Flat antenna arrays, particularly those with planar configurations, can be vulnerable to high-power jamming signals. Jamming signals can interfere with the operation of the antenna array, affecting its ability to detect and track targets accurately. Additional measures, such as anti-jamming techniques or adaptive signal processing, may be necessary to mitigate the effects of jamming.

4. Simultaneous functionality: While multi-beam systems can track multiple targets, perform different functions, or communicate with multiple satellites, it is important to note that these functions may not be truly simultaneous. Switching between beams or temporarily deactivating previous functions may be required to allocate resources effectively. The

system's capabilities depend on factors such as the number of beams, processing capabilities, and the complexity of the tasks involved.

Addressing these challenges requires careful system design and engineering considerations. Tradeoffs must be made to optimize performance based on specific application requirements and constraints. Ongoing research and development directional antenna arrays aim to overcome these challenges and improve the capabilities of multi-beam antenna systems.

It is obvious that scanning beams in phased array systems can lead to shorter illuminating time for each satellite and can limit the link budget for satellite communication. The time spent on actively communicating with each satellite is reduced as the scanning beam moves quickly across different targets or satellites. This can result in a decrease in the amount of information received from each target. The link budget is a measure of the overall communication performance, taking into account factors such as transmit power, receiver sensitivity, path loss, and interference. Switching the beam rapidly between multiple satellites can lead to a decrease in the link budget for each individual satellite, as the time allocated for communication with each satellite is reduced (Figure 2.12). The same is true for phase arrays for radar applications. Scanning beam dramatically decreases the illuminating time for targets. Receiving information from targets decreases as the number of detected targets increases. Omnidirectional (isotropic) antenna elements in a planar phase array limit the covered space sector.

Multi-beam phased antenna arrays face challenges in terms of scanning time and the impact on communication or radar illumination time. Scanning a narrow beam with high direction accuracy requires scanning through multiple positions, which can take a significant amount of time. This can result in shorter communication time or illumination time for each satellite or target.

Moreover, the number of beams in a multi-beam antenna system can impact the gain and sensitivity of each individual beam. As the number of beams increases, the aperture is divided among the beams, leading to a decrease in gain for each beam. This reduction in gain directly affects the sensitivity and range of the antenna array.

In addition, switching antenna elements or beams for observing different space sectors or enabling multi-band functions can further impact the time allocated for communication with each satellite or the link budget for radar applications. The need for beam switching or element switching introduces additional overhead and can reduce the overall communication time or radar illumination time for individual targets (Figure 2.13).

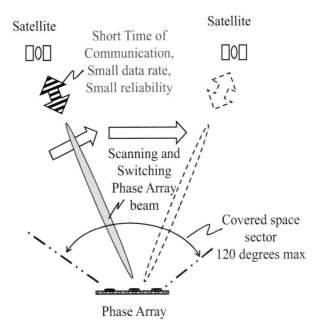

Figure 2.12 Scanning phase array beam in a phase array limiting the satellite link budget. Switching beam between multiple satellites makes the link budget dramatically worse.

Electronically steerable panel antennas, such as phased array antennas, have limitations when it comes to frequency band and bandwidth. The steering capability of a phased array is directly related to the frequency being used, and it becomes challenging to achieve wideband or multi-frequency operation with a single phased array.

There is indeed a tradeoff between frequency band, gain, and switching time in phased arrays. Increasing the frequency band or data rate may require sacrificing some of the gain or increasing the switching time, which can impact the overall signal quality and data transmission capabilities.

In contrast, using an array of directional antennas can offer more flexibility in terms of frequency band and bandwidth. Each individual directional antenna can be designed for a specific frequency, allowing for multi-frequency operation within the array. By directly digitizing and synchronizing the signals at each antenna element, the array can be distributed on concave/convex surfaces, around the perimeter of a vehicle, or among a swarm of manned or unmanned vehicles.

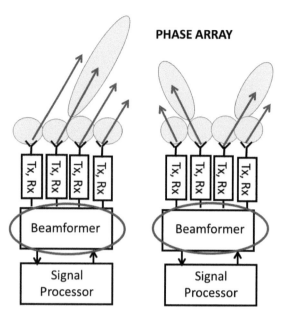

Figure 2.13 A change in the number of beams in a phase array for the same antenna array aperture leads to decreasing gains in each beam and correspondingly decreasing sensitivity and range of the antenna array.

This approach enables more efficient utilization of the available frequency spectrum and allows for better adaptation to different communication requirements. Each directional antenna within the array can be optimized for its specific frequency band, resulting in improved performance and flexibility compared to a single phased array covering a wide frequency range. By leveraging an array of directional antennas, it becomes possible to overcome some of the limitations associated with frequency band and bandwidth in electronically steerable panel antennas.

2.5 MIMO Antenna Arrays

"Multiple-input and multiple-output (MIMO) is a method for multiplying the capacity of a radio link using multiple transmission and receiving antennas to exploit multipath propagation" [5–7]. For wireless systems, MIMO antenna arrays normally use multiple antennas for transmitting and receiving. MIMO can be referred to as a class of techniques for sending and receiving more than one data signal simultaneously over the same radio channel by exploiting multipath propagation. Modern MIMO (Figures 2.14 and 2.15) can be used for multiple data signals sent to different receivers. MIMO radars can be referred to as phased array radar employing digital receivers and waveform

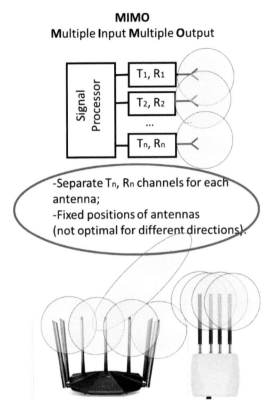

Figure 2.14 Orthogonal separation of transmitting and receiving signals allows to improve spatial resolution and to arrange antennas in a smaller volume. However, fixed positions of antennas are not optimal for beamforming. As a result, antenna arrays lose gain and sensitivity/range.

generators distributed across the aperture. MIMO radar signals propagate in a fashion similar to multistatic radars. However, instead of distributing the radar elements throughout the surveillance area, antennas are closely located. In a traditional phased array system, additional antennas and related hardware are needed to improve spatial resolution. MIMO radar systems transmit mutually orthogonal signals from multiple transmit antennas, and these waveforms can be extracted from each of the receiving antennas by a set of matched filters [5–7]. MIMO antenna arrays may not provide the same level of gain, sensitivity, or range as phased arrays or arrays of directional antennas with the same number of antennas. This is because the positioning of multiple omnidirectional antennas in a MIMO array may not be optimized for different communication or radar beam directions. The design of a MIMO antenna array involves a tradeoff between choosing the best direction for

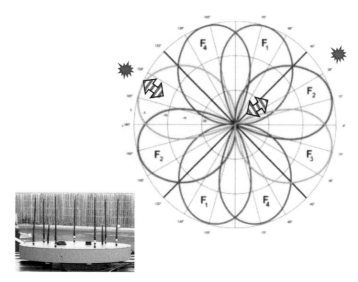

Figure 2.15   Diagram of a MIMO switching beam with an eight-element antenna array and antenna patterns.

channel communication and maximizing gain. In summary, MIMO antenna arrays offer advantages in terms of increased capacity and improved signal quality by exploiting multipath propagation. However, they may not provide the same level of gain and directionality as phased arrays or arrays of directional antennas, and their performance is influenced by the specific application requirements and tradeoffs.

Today, in complex intelligence, surveillance, and reconnaissance (ISR) applications, there is often a need to receive continuous intelligence information from a 360° × 360° area and from multiple targets simultaneously. Traditional scanning or switching beam antenna arrays can introduce interruptions in the receiving process, leading to a decrease in the time of illuminating or receiving signals from the targets.

When using a scanning or switching beam antenna array, the time spent on each target is reduced as the antenna switches between different targets. For example, if there are eight antennas and two targets, the illumination time for each target is reduced by a factor of 8, and for two targets, it is reduced by a factor of 16.

This limitation can impact the effectiveness of the ISR system, as it may result in reduced data collection and slower response times. To overcome this challenge, alternative antenna array configurations, such as multi-beam or multi-channel arrays, are often employed. These configurations allow for

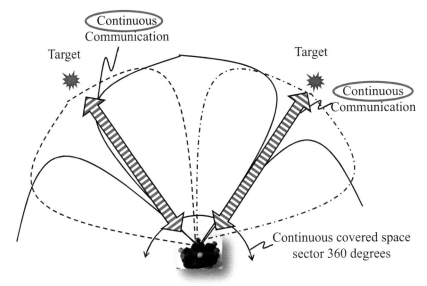

Continuous Communication

Target

Target

Continuous Communication

Continuous covered space
sector 360 degrees

Array of Directional Antennas

Figure 2.16 Array of directional antennas with staring beams.

simultaneous reception from multiple targets, minimizing interruptions and maximizing the time of illumination for each target.

Directional antenna arrays with staring beams have the advantage of providing simultaneous and continuous illuminating and receiving signals from multiple targets or satellites. Unlike scanning or switching beam arrays, directional antenna arrays cover the entire area of observation simultaneously, providing the same gain for all directions regardless of the number of targets.

By maintaining continuous communication with all satellites or continuous illumination of all targets, directional antenna arrays can increase the range and reliability of the system. This is especially beneficial in applications such as intelligence, surveillance, and reconnaissance (ISR), where real-time and continuous data collection is crucial. The use of directional antenna arrays allows for efficient and effective coverage of the entire observation area without the need for scanning or switching beams. This ensures that all targets or satellites within the coverage area are constantly illuminated and can be tracked or communicated by simultaneously improving the overall performance of the system.

An important tradeoff in surveillance radar systems is the balance between illumination time (time on target), update rate, directional accuracy, and velocity resolution. These factors are influenced by various antenna

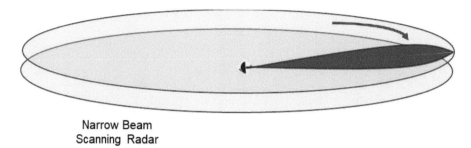

Narrow Beam
Scanning Radar

Figure 2.17 Mechanical scanning narrow (pencil) beam antenna provides 360° by an azimuth area of observation.

parameters, including beamwidth, clutter rejection capability, and antenna gain. A longer illumination time allows for better Doppler analysis and improves the signal-to-noise ratio, which is beneficial for target detection and tracking. However, a longer illumination time may result in a slower update rate, which could sometimes impact the system's ability to provide real-time information to the user.

The antenna beamwidth plays a role in this tradeoff. A wider beam covers a larger area and captures more information. However, a wider beam also increases the clutter level, which can affect target detection in the presence of background noise.

The power budget is another factor to consider. A wider beam typically leads to lower antenna gain, which affects the system's ability to detect weak signals. However, a wider beam can also provide higher coherent integration gain, which improves the signal-to-noise ratio over a longer integration time. Finding the right balance between these factors is crucial in designing a surveillance radar system that meets the requirements of the user. It involves considering the specific operational needs, such as the desired update rate, Doppler analysis capabilities, clutter rejection, and power budget, and optimizing the antenna parameters accordingly [8, 9].

2.6 Evolution of Antenna Array Systems

Mechanical scanning antennas and antenna arrays have been used to scan wide areas of observation. By using a narrow, focused beam, these antennas can achieve high gain and maximize radar range. The mechanical scanning system allows for high directional accuracy, which is important for target tracking and surveillance.

However, one limitation of mechanical scanning is that it requires a relatively long time to cover the full area of observation. This is because the

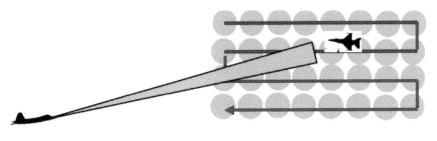

**Narrow Beam
Scanning Radar
AESA**

Figure 2.18 Electronically scanning narrow (pencil) beam antenna array provides high directional accuracy and a very fast (0.5–1 seconds) scanning time for a limited area of observation.

antenna or array needs to physically move to scan different directions. The scanning process is sequential, which means that it may not provide real-time or continuous coverage of the entire area. This limitation can impact the system's ability to detect fast-moving targets or provide timely updates to the user. Despite this drawback, mechanical scanning antennas have been widely used in radar applications, especially when the need for high gain, long-range coverage, and precise directional accuracy outweighs the requirement for real-time and continuous scanning. The choice between mechanical scanning and other scanning methods depends on the specific application, operational needs, and tradeoffs between scanning time, coverage area, and system requirements.

Electronically scanning antenna arrays dramatically decrease the time for observation. But the application of phase array panels limits the area of observation (Figure 2.18). Multi-beam, multi-function digital scanning antenna arrays offer several advantages for various applications such as target detection, simultaneous communication with multiple customers, and navigation. These systems provide flexibility and versatility by electronically steering multiple beams to different directions without the need for mechanical movement. However, it is important to note that the performance of digital scanning antenna arrays can be influenced by certain limitations. One limitation is the frequency band coverage. Phase antenna arrays are typically designed for a specific frequency band, and extending the bandwidth can be challenging due to the inherent constraints of phase control and antenna element spacing.

Additionally, the angle of view or coverage area of planar one-sided phase antenna arrays is limited. The directional characteristics and beamwidth

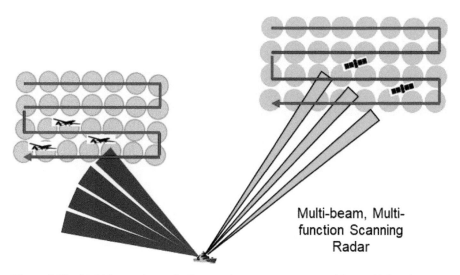

Figure 2.19   Multi-beam electronically scanning antenna array provides multi-function pos-sibilities in a few space sectors.

of each beam are determined by the antenna design, and there are practical limitations to the achievable angle of view. Wide-angle coverage may require complex designs or additional antenna elements, which can impact the over-all performance and system complexity (Figure 2.19).

Therefore, when designing and implementing digital scanning antenna arrays, it is crucial to carefully consider the specific application requirements, such as the desired frequency band, angle of view, and tradeoffs between sys-tem complexity, beam steering capabilities, and performance.

Holographic radars indeed represent a significant advancement in the antenna array design, transitioning from scanning antenna arrays to staring antenna arrays. By implementing holographic principles, these radars pro-vide continuous coverage of the entire surveillance volume, enabling pro-longed dwell time on each signal source or target. This continuous coverage allows for uninterrupted communication links and real-time detection and tracking of targets with high accuracy (Figure 2.20).

One of the key advantages of holographic radar is its ability to virtu-ally scan the entire space without physically moving the antenna. This vir-tual scanning enables a higher update rate than traditional scanning antenna arrays, providing more timely and accurate information. The radar system records a real-time hologram that captures fully sampled amplitude and phase information from every object within the volume. This hologram can then be

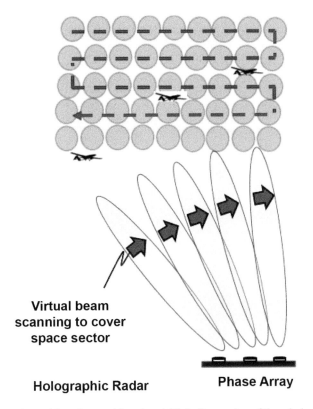

**Virtual beam
scanning to cover
space sector**

Holographic Radar   **Phase Array**

Figure 2.20   Holographic radar provides virtual (digital) scanning of the whole area of observation and wide possibilities for digital signal processing.

processed to extract range, azimuth, elevation, and Doppler information for each detected object.

Holographic radar systems also offer improved clutter rejection capabilities. Advanced tracking algorithms can discriminate between moving targets and clutter, allowing for effective clutter removal without sacrificing sensitivity. This enables more accurate target detection and tracking in complex environments [10].

In terms of system architecture, holographic radar systems can be less complex compared to conventional surveillance array radar architectures. However, their capabilities can surpass even the most sophisticated active electronically scanned array (AESA) radars. This combination of lower complexity and superior performance positions holographic radars as potential challengers to conventional radars in terms of both cost and performance [11]. Overall, holographic radars represent a significant leap forward in antenna

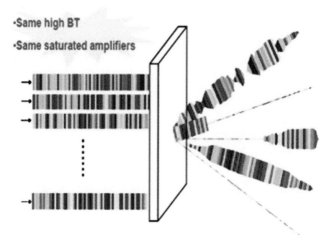

Figure 2.21   Sample of space–time coding, generic case [8, 12].

array technology, offering continuous coverage, high update rates, accurate target detection, and improved clutter rejection capabilities. Their potential to challenge conventional radars in cost and performance makes them an exciting area of development in the field of radar systems.

Space–time coding is a technique that can be utilized to enhance the recovery of angular separation in signals transmitted in different directions. This method involves the simultaneous transmission of different signals in various directions, effectively coding both space and time. On the receiving end, these signals are processed coherently and in parallel, enabling their separation (Figure 2.21).

The utilization of space–time coding offers several advantages. One key benefit is improved target extraction, particularly for slow-moving targets, amidst challenging conditions such as clutter, multipath interference, and noise. By exploiting longer observation times and potentially wider bandwidths, space–time coding enables a more effective identification of targets.

In a typical configuration, as depicted in Figure 2.6, different codes are transmitted through individual antenna elements or sub-arrays. It is preferable for these codes to have constant amplitude, as it enhances the efficiency of the amplifiers. As a result, the antenna transmits a modulated signal that exhibits variations in different directions, corresponding to the different codes employed.

Overall, the space–time coding is a valuable technique that combines space and time to improve the recovery of angular separation in transmitted signals. By leveraging the advantages of this method, such as enhanced target

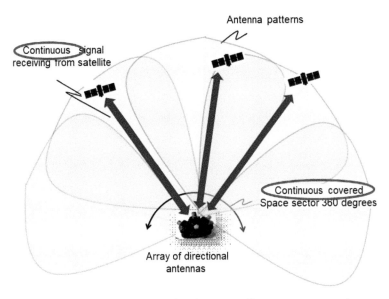

Figure 2.22   Array of directional antennas (fly eye antenna array).

extraction and improved identification, it becomes possible to overcome the challenges associated with clutter, multipath interference, and noise [8, 12].

An array of directional antennas can cover all required space, providing high gain in any direction for multiple beams and non-interrupting illumination of all targets or communication with multiple satellites with non-interrupting full link budgets (Figure 2.22) [13–16].

All space antenna arrays are typically composed of multiple antenna elements or directional antennas arranged in a specific pattern to form directional radiation characteristics. These elements or antennas can be positioned in a variety of ways to achieve the desired beamforming and radiation properties. One common approach is to position the antenna elements or directional antennas within a dielectric substrate, which can have a constant or variable dielectric constant. This arrangement allows for control over the propagation of electromagnetic waves and can help shape the antenna's radiation pattern.

Alternatively, a three-dimensional metamaterial substrate can be used to achieve the desired radiation characteristics. Metamaterials are engineered materials with unique electromagnetic properties, which can be utilized to manipulate the propagation of waves and achieve specific antenna performance. In the case of multi-beam antenna arrays, the arrangement of antenna elements can involve positioning them inside and on the surface of a dielectric

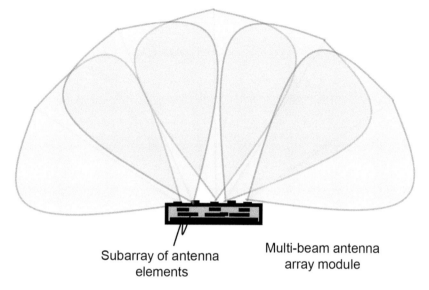

Subarray of antenna
elements

Multi-beam antenna
array module

Figure 2.23 Transmitter antenna module (TAM).

substrate, as illustrated in Figure 2.23. This configuration allows for the generation of multiple beams with different directions and beamwidths.

Additionally, directional antennas can be arranged inside a dielectric material layer that is attached to the main transceiver substrate. This approach provides further flexibility in designing the antenna array and achieving the desired radiation properties.

In summary, the arrangement of antenna elements or directional antennas within dielectric substrates or metamaterial structures offers control over the radiation characteristics and enables the formation of multi-beam antenna arrays or other desired antenna configurations.

In multi-beam antenna arrays, a plurality of sub-array modules can be used to cover the entire space or area of observation simultaneously. These sub-array modules consist of directional antennas with overlapping antenna patterns in one or more directions, allowing for the creation of multi-axis monopulse sub-arrays. This configuration enables the antenna array to have enhanced angular resolution and tracking capabilities.

Antenna elements within the array can be arranged to operate as resonance antennas, designed for one or a few specific frequencies. Examples of resonance antennas include Yagi antennas and helical antennas, which are known for their narrowband operation and high gain at specific frequencies.

To extend the bandwidth of the antenna array and achieve frequency independence, non-resonant frequency traveling-wave antennas can be

FLY EYE ANTENNA ARRAY

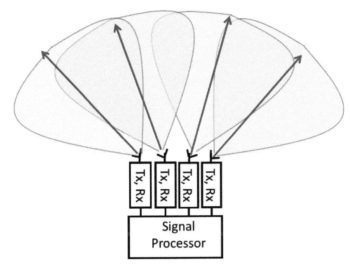

Figure 2.24 Directional antennas provide better interference protection between separate channels.

employed. Traveling-wave antennas can be arranged in two main forms: discrete radiators placed along an axis at a specific distance from each other and continuous radiators that extend in the direction of the axis. These antennas allow for broader frequency coverage and can be designed to operate over a wide range of frequencies.

The selection of resonance or non-resonance antennas depends on the specific application requirements, including the desired frequency range, bandwidth, and gain characteristics. By combining different types of antennas and arranging them in a suitable configuration, multi-beam antenna arrays can achieve simultaneous coverage of multiple frequencies and enhance the overall performance of the system (Figure 2.24).

The multi-axis distribution of overlapping antennas in an array enables the reception of maximum information from multiple sources or objects within the coverage area. By coupling each directional antenna with separate transmitting/receiver channels and utilizing direct digitization techniques, it becomes possible to receive a digital hologram that describes the entire 3D object or scene simultaneously.

Compared to scanning systems, which require sequential scanning of different directions to gather information, multi-axis antenna arrays with direct digitization offer significant advantages in terms of speed and efficiency. The parallel processing capability of the array allows for faster acquisition of data, as multiple channels can simultaneously capture information from different directions. This not only improves the update rate but also enables real-time or near-real-time analysis and processing of the received data.

The digital hologram obtained from the multi-axis antenna array provides a detailed representation of the 3D object, including its range, azimuth, elevation, and other characteristics. This information can be further processed using advanced algorithms and techniques to extract valuable features, track moving objects, perform target identification, and enhance overall situational awareness.

The combination of multi-axis distribution, separate channels, and direct digitization in antenna arrays facilitates the rapid acquisition of comprehensive 3D information, making them suitable for a wide range of applications, including surveillance, imaging, radar, and communication systems.

All-space antenna array with a staring configuration, composed of a multitude of directional antennas, has the capability to cover the entire sky and enable simultaneous multi-channel, multi-band, and multi-function operations. This type of antenna array offers several advantages for communication systems. By utilizing a staring configuration, the antenna array can maintain continuous illumination of multiple objects or sources without interruption. This allows for reliable and non-interrupting communication with multiple targets simultaneously. The continuous illumination ensures a consistent link budget, which refers to the overall performance and strength of the communication link.

The use of directional antennas in the array enables increased gain and sensitivity, resulting in an enhanced link budget. Directional antennas focus their radiation in specific directions, allowing for better reception and transmission of signals. This increased gain improves the signal strength and extends the range of the communication system.

Simultaneous multi-channel operation enables the antenna array to handle multiple signals or connections at the same time, providing efficient communication with multiple sources. The multi-band capability allows for operation across different frequency bands, accommodating various communication requirements.

The all-space antenna array with a staring configuration offers continuous and reliable illumination of multiple objects, increased link budget, and the ability to support simultaneous multi-channel, multi-band, and multi-function regimes. These features make it well-suited for applications requiring robust and efficient communication in complex environments (Figure 2.25).

Figure 2.25 Prototype of fly eye antenna array inspired by nature.

References

[1] Federal Standard 1037C: Glossary of Telecommunications Terms. US General Services Administration. pp. O-3. ISBN 1461732328. National Telecommunication Information Administration, 1997.

[2] Oscar Liang web page. How Antenna Gain affects Range in FPV - Oscar Liang

[3] Wikipedia, Phase Array.

[4] Alison K. Brown and Ben Mathews, "GPS Multipath Mitigation Using a Three Dimensional Phased Array" *NAVSYS Corporation, 2015.*

[5] Lipfert, Hermann, "MIMO OFDM Space Time Coding – Spatial Multiplexing, Increasing Performance and Spectral Efficiency in Wireless Systems", Part I Technical Basis (Technical report). Institut für Rundfunktechnik, August 2007.

[6] Kaboutari, Keivan, Hosseini, Vahid "A compact 4-element printed planar MIMO antenna system with isolation enhancement for ISM band operation". AEU - International Journal of Electronics and Communications. 134: 153687. doi:10.1016/j.aeue.2021.

[7] Emil Bjornson, "Cellular Multi-User MIMO: A Technology Whose Time has Come" | Wireless Future Blog (ellintech.se), NOVEMBER 11, 2016.

[8] F. Le Chevalier, Wideband wide beam motion sensing, in: J. Taylor (Ed.), Advanced Ultrawideband Radar: Targets, Signals and Applications, CRC Press, 2016 (Chapter 12).

[9] M. Cherniakov (Ed.), Bistatic Radar: Principles and Practice, Wiley, Chichester, UK, 2007.

[10] Gary Kemp, Holographic radar brings a new dimension to sensing and instrumentation on T&E ranges Collision avoidance, wind farms

and scoring, NDIA test and evaluation conference, S4923-P-069 v0.2, March 2011.

[11] Stephen Harman, Aveillant Ltd., Cambridge, U.K. Holographic Radar Development | 2021-02-07 (microwavejournal.com), www.microwave-journal.com/articles/35410-holographic-radar-development

[12] François Le Chevalier *, Nikita Petrov Delft University of Technology, The Netherlands URSI-France 2018 Workshop: Geolocation and navigation / Journées URSI-France 2018.

[13] Pavlo A. Molchanov, AMPAC Inc.; V.M. Contarino, R Cubed Inc.; O.V. Asmolova, AETHER Inc. Protected GPS Directional Antenna Array. JSDE/ION Joint Navigation Conference 2012, Session C7.

[14] Pavlo A. Molchanov, A. Gorwara, "Fly Eye radar concept". IRS2017. International Radar Symposium, Prague, July 2017.

[15] Pavlo A. Molchanov "Multi-beam multi-band antenna array module" US patent appl. 17/971,616, 10/23/2022.

[16] Pavlo A. Molchanov, Olha V. Asmolova, "All-digital radar architecture" AETHER Inc. (PDF) Paper 9248-11, Security+Defense Conference, Amsterdam, September 25, 2014 (researchgate.net).

3

Recording of a Real-time Digital Hologram

3.1 Holographic Radar

The term "holography" comes from the Greek words "holos," meaning "whole" or "entire," and "grafe," meaning "writing". At first, the term "holography" was introduced by Dennis Gabor in 1948 in "A new microscopic principle" [1]. The name was chosen to indicate that the method records the entire field information (i.e., amplitude and phase) and not just the usual intensity. Initially, Gabor proposed this technique to "read" optically electron micrographs that suffered from severe spherical aberrations. Gabor was awarded the Nobel Prize in Physics "for his invention and development of the holographic method" in 1971. The proof of principle demonstration was performed entirely in the optical domain, and holography has since remained largely connected with optical fields. According to D. Gabor, holography is a two-step process:

1. Writing the hologram, which involves recording on film the amplitude and phase information.

2. Reading the hologram, by which the hologram is illuminated with a reference field similar to that in step 1 [1–3].

The recording of the entire field information in the radio-frequency (RF) domain is called RF holography. RF holography methods have been applied for holographic detection and for imaging radars. A holographic radar implements Skolnik's vision of ubiquitous radar, meaning that a holographic radar is continuously staring (rather than scanning) at a whole volume of space. It is recording fully sampled amplitude, phase, and frequency information from every object within the volume and provides range, azimuth, elevation, and Doppler information for every detected object. Rather than scanning a radar beam across an entire observation area, a holographic radar employs a staring array to cover the whole area, all the time. This allows tracking the entire object or multiple small targets simultaneously over the whole area of

53

observation. Holographic data processing and recording of wave fronts or digital hologram was first proposed in holographic radar [2, 3].

Holographic radar, based on the principles of holography, is a radar system that continuously stares at the entire coverage volume, providing complete dwell time on each signal source or target. It records the fully sampled amplitude, phase, and frequency information from every object within the volume, enabling range, azimuth, elevation, and Doppler information for each detected object.

Unlike traditional scanning radar systems that sweep a radar beam across the observation area, a holographic radar utilizes a staring array to cover the entire area simultaneously. This allows for the continuous detection and tracking of multiple targets, including small and low-profile objects such as drones, with high accuracy and in real time.

The architecture of holographic radar is generally less complex compared to conventional surveillance array radar systems. However, holographic radars offer capabilities that surpass even the most sophisticated active electronically scanned array (AESA) radars. By leveraging the principles of holography, a holographic radar achieves a comprehensive and continuous view of the surveillance volume, providing enhanced situational awareness and improved detection and tracking capabilities.

The use of holographic data processing and recording wave front or digital holograms is a key feature of the holographic radar. This approach allows for the recording and analysis of the complete field information, including amplitude, phase, and frequency. It enables advanced signal processing techniques and provides a rich dataset for accurate and detailed target analysis. Holographic radar offers significant advantages in terms of continuous coverage, high update rates, accurate target tracking, and simplified architecture. It represents a powerful and promising technology for surveillance, detection, and tracking applications in both civilian and military domains.

In a holographic radar, the transmitting signal can cover multiple targets simultaneously. The receiver then virtually scans the entire area of observation, recording the wave front or digital hologram of the received signals. By digitizing the incoming wave fronts using an array of antennas, the direction finder in the radar system monitors the wave fronts across the aperture and produces a radio-frequency hologram. This allows for the numerical reconstruction of the hologram and enables direction finding in the radio-frequency domain [2, 3].

Unlike traditional radar systems, a holographic radar focuses on measuring and recording the phase across the aperture rather than using photographic emulsion or intensity-based measurements. By sampling the phase

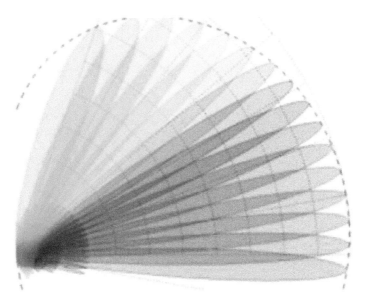

Figure 3.1 Wave front of interfering electromagnetic waves that reflected from object recorded as digital hologram with reference to transmitted signals or digitizing/sampling signals [7].

at discrete points within the aperture using closely spaced antenna elements, the direction of arrival can be determined with high accuracy, equivalent to recording the holographic image of a distant point source.

The holographic radar offers several advantages in terms of information extraction and target detection. By dividing the radar return into narrower components, the radar can retain and report fine-grain Doppler signatures, providing detailed information about target positions, velocities, and trajectory changes. This enables effective tracking, threat assessment, and discrimination between moving targets and clutter.

The flexibility and utility of holographic radar are further enhanced by its ability to rapidly determine changes in trajectory and provide continuous surveillance and tracking updates. With highly flexible radar coverage and the capability to detect and monitor targets with fine-grain resolution, the holographic radar offers a solution for ubiquitous radar sensing, meeting the demands of diverse applications and environments.

The tracking signal processing algorithms employed in holographic radar play a crucial role in discriminating moving targets from clutter and enhancing the radar's detection and tracking capabilities. These algorithms enable efficient target tracking, trajectory analysis, and threat

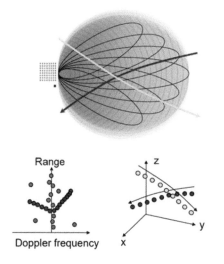
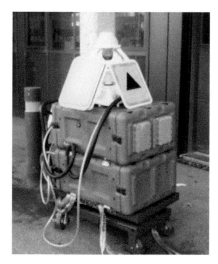

Figure 3.2   Holographic radar recording fully sampled amplitude and phase and frequency information in the whole area of observation [5].

assessment based on the rich information provided by holographic radar technology (Figures 3.2, 3.3) [4, 5].

Staring holographic radars typically have a more simplified hardware structure in terms of RF (radio frequency), IF (intermediate frequency), and control components compared to scanning radars. They do not require any moving parts for scanning the radar beam. However, this simplicity in hardware is compensated by increased processing complexity. Staring radars have the capability to process multiple beams simultaneously, unlike scanning radars that typically process one beam at a time. This means that the amount of data processed in a single dwell, or observation period, is generally higher in staring radars. This increased data processing is necessary to achieve similar sensitivity to an equivalent scanning radar.

Since staring radars rely on digital holography and beamforming techniques, they can take advantage of high-resolution Doppler processing and other benefits. However, processing a larger amount of data and performing complex signal processing tasks, such as fast Fourier transforms (FFT), increases the computational load on the radar's signal processing system.

The processing load of a fast Fourier transform increases with the length of the data being processed. As a result, staring radars may require more powerful and efficient signal processors to handle the increased computational demands. This higher processing complexity is necessary to extract detailed information from the received signals and achieve the desired radar performance. Staring holographic radars offer advantages such as simplified hardware and simultaneous processing of multiple beams [6–8].

The main difference between scanning antenna systems and holographic systems is in terms of coverage and recording of signals. Scanning antenna systems with narrow beams have limited coverage and can only record signals within their narrow beamwidth. They typically capture direct reflected signals, but signals reflected from outside the antenna's angle of view are not captured. This can result in a loss of information about the object or the communication signals.

In contrast, holographic systems, whether optical or digital, aim to record the wave front of signals reflected from the entire object or area of observation. This is achieved by using an "object beam" that captures the signals reflected from the entire object, in conjunction with a "reference beam" for comparison. The recording process of a true hologram involves capturing the complete wave front information.

When scanning systems are used for holography, they can be considered quasi-holograms because they do not capture the full wave front information from the entire object. The limited coverage of the scanning antenna's narrow beam pattern results in a loss of information about the object or the communication signals. This can lead to a decrease in data rate and link budget, as the missing information cannot be recovered.

Therefore, for applications where capturing the complete wave front information is crucial, such as in holography, a scanning antenna system may not be sufficient. Holographic systems, with their ability to simultaneously cover the entire object or area of observation, provide a more comprehensive and accurate representation of the signals, leading to higher data rates and improved link budget.

3.2 Near-field Holographic Imaging

Historically, holograms have been primarily used for recording and recreating 3D optical images. However, the concept of holography can also be extended to RF signals for applications such as radar systems.

In RF holography, the goal is to record and recreate 3D images of objects or perform 3D imaging and identification of multiple targets in radar systems. To achieve this, the radar system must capture the near-field wave front of the reflected RF signals and process not only the amplitudes but also the phase- and frequency-shifted near-field components.

When an object is illuminated with coherent continuous-wave (CW) electromagnetic signals, the reflected signals in the far field can exhibit

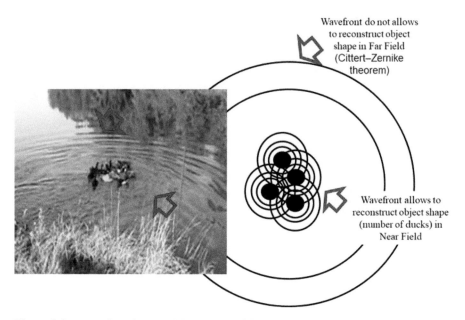

Figure 3.3   Wave front in near field consists of frequency components, allowing to reconstruct the object shape. Wave front in far field consists of only coherent waves and does not allow to reconstruct the object shape.

coherence and have the same frequency and wavelength, as described by the Cittert–Zernike theorem (CZT). However, in the far field, the wave fronts alone do not allow for the direct reconstruction of the object shape. In order to reconstruct the 3D shape of the object from RF holograms, additional information and processing techniques are required. This can involve complex algorithms and signal processing methods to extract the necessary spatial and geometric information from the holographic data. By analyzing the phase- and frequency-shifted components of the near-field wave fronts, along with amplitude information, it is possible to reconstruct a 3D representation of the object or perform imaging and identification of multiple targets in the radar system (Figure 3.3).

Wave fronts in the near field of an object carry valuable information about the reflected waves, including time and phase shifts. This information can be utilized to reconstruct the shape and size of the object. In addition to time and phase shifts, the dielectric properties of the object can introduce further phase shifts proportional to the object's dielectric coefficient. This phase shift, similar to the concept of impedance in the Smith chart, provides information that can aid in identifying the object. By transforming the reflected signals from the time–space domain to the frequency domain,

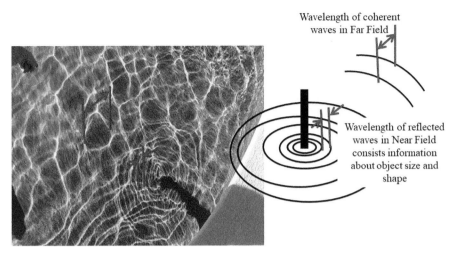

Figure 3.4 Waves reflected from pool walls are coherent in far field. Waves reflected from objects with a smaller wavelength size comprise frequency components corresponding to object size and shape if in the near field.

it becomes possible to analyze the frequency components that contain information about the object's shape, size, and dielectric properties. This can be achieved through techniques such as Fourier analysis or other signal processing methods.

It is important to note that in the near field, the dispersion of continuous waves can create additional frequency components that correspond to the object's size and shape. These frequency components are not necessarily harmonics of the transmitting frequency but rather arise due to the complex interaction between the electromagnetic waves and the object in the near field. These frequency components are typically found within a certain range, such as 5%–50% of the transmitting frequency.

By analyzing the frequency components and their characteristics, such as the amplitude and phase, it is possible to create a spectrum signature that represents the object and contains information about its shape, size, and dielectric properties. Analysis of wave fronts in the near field, along with the transformation to the frequency domain, allows for the extraction of valuable information about the object being observed, enabling the reconstruction of its shape, identification, and characterization based on its frequency spectrum signature (Figure 3.4).

In order to record a digital hologram, the antenna array is typically arranged as an array of directional antennas that cover the area of observation

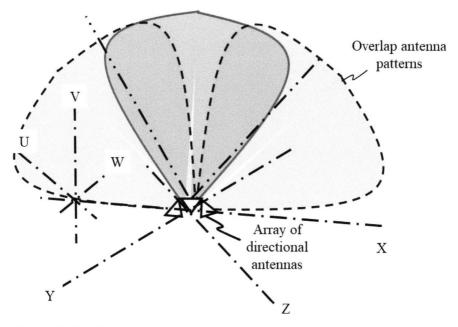

Figure 3.5   Directional antenna array with overlapping antenna patterns in a few axes providing more accurate information about the object and a better image resolution.

or the area where the objects of interest are located. The antenna patterns are designed to overlap, at least in one axis, creating antenna sub-arrays.

The concept of overlapping antenna patterns, also known as the fly eye antenna array or monopulse directional antenna array, has been employed in the design of antenna arrays for recording digital holograms using RF electromagnetic waves. The overlapping patterns provide several advantages, including improved image resolution and the ability to capture additional information about the object being observed. As shown in Figure 3.5, antenna patterns are overlapping in axis *X*, but patterns can be overlapping in axes *Y*, *U*, and *W*. By overlapping the antenna patterns in one or more axes, multi-axis overlapping antennas can offer a better image resolution, allowing for more detailed reconstructions of the object's shape and features. This can lead to enhanced imaging capabilities and a higher level of information extraction.

Furthermore, the monopulse method of direction finding, which involves comparing the signals received by overlapping antennas, can provide improved accuracy in determining the direction of arrival of the electromagnetic waves. This method can also contribute to higher-resolution imaging. The use of multi-axis overlapping antennas in an antenna array for recording digital holograms offers advantages in terms of image resolution, information extraction, and direction finding accuracy [8–15].

Figure 3.6 illustrates a diagram of a system for recording a near-field digital hologram using a continuous-wave radio-frequency electromagnetic transmission. The system is designed to cover the entire object or the area of observation, ensuring that the antenna or antenna array captures the complete wave front of the reflected electromagnetic signals. In this setup, the transmitted continuous-wave RF signal interacts with the object, and the reflected (and possibly partially penetrated) electromagnetic signals are received by one or an array of antennas positioned in the near-field area. It is crucial for a hologram recording that the antenna or antenna array covers the entire object or the area of observation to capture the complete information.

The received signals carry three components that provide information about the object:

1.   Time shift: The signals experience time delays from different points on the object, which encode information about the object's shape.

2.   Phase shift: The signals exhibit phase shifts (and corresponding time shifts) due to the arrival of reflected signals from different directions. This phase shift provides directional information.

3.   Amplitude and time shift: Some of the reflected signals penetrate into the object's body or volume, leading to changes in amplitude and time shift. This component contains information about the object's impedance and the properties of the surrounding medium.

The signals received by the antenna or antenna array, which include the time shifts, phase shifts, and amplitude variations, are combined and can be directly digitized and recorded as a digital hologram. The processor time serves as the reference signal for the hologram recording process. Alternatively, if available, stable sampling signals or synchronization signals with smaller phase errors can be used as references for the received signals.

By recording the complete near-field wave front information of the reflected signals, the digital hologram captures the three-dimensional information about the object's shape, size, and impedance characteristics, as well as information about the surrounding medium. Overall, the system depicted in Figure 3.6 enables the recording of a near-field digital hologram, providing comprehensive information about the object under observation.

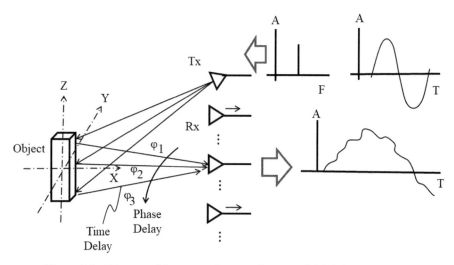

Figure 3.6 Diagram of the system for recording near-field digital holograms.

Figure 3.7 shows the signal processing stage in which the recorded digital holographic data is transformed into the frequency domain. This transformation allows for the extraction and manipulation of frequency components that carry information about the shape, size, and dielectric properties of the object and the surrounding medium.

In the frequency domain, the digital holographic data is analyzed and processed to group, filter, and code the relevant frequency components. These components correspond to the various characteristics of the object, such as its shape, size, and dielectric properties.

Once the frequency components are identified and separated, they can be visualized on a display in the form of a "Waterfall" plot. This real-time dynamic image provides a comprehensive view of the frequency components and their variations over time. By utilizing the frequency domain representation of the digital holographic data, it becomes possible to visualize and analyze the object's characteristics in a more intuitive and informative manner.

The "Waterfall" display allows for real-time monitoring and observation of the object, enabling further analysis and interpretation of its properties. Figure 3.7 illustrates the stage where the digital holographic data is transformed into the frequency domain, facilitating the visualization and analysis of frequency components that contain valuable information about the object's shape, size, and dielectric properties.

Figure 3.8 represents the final stage of the holographic radar system, where the frequency components obtained from the processing stage are

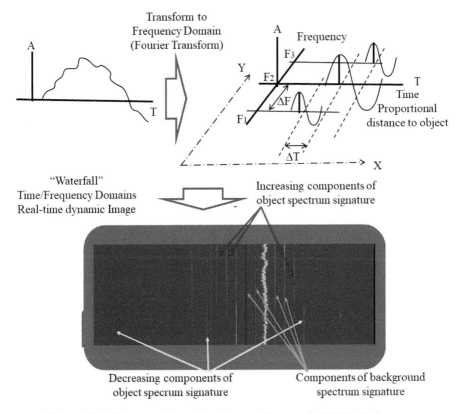

Figure 3.7 Diagram of the system for recording near-field digital holograms.

transferred back to the time domain. This can be achieved, for example, by performing an inverse Fourier transform on the frequency components. By converting the frequency components back to the time domain, the holographic radar system can reconstruct the object in the time–frequency–space domains. The reconstructed object can be visualized and analyzed in two or multi-axis coordinates, providing a comprehensive representation of its characteristics.

It is important to note that for real digital hologram recording, the recording antenna array or multiple antenna arrays must simultaneously cover the entire object or the area of observation. Scanning antenna systems with narrow beams that do not cover the entire object at once can only record direct reflected signals and lose signals reflected outside the antenna's angle of view.

The main condition for recording real digital holograms is the recording of the wave front of signals reflected from all objects (object beam) relative to

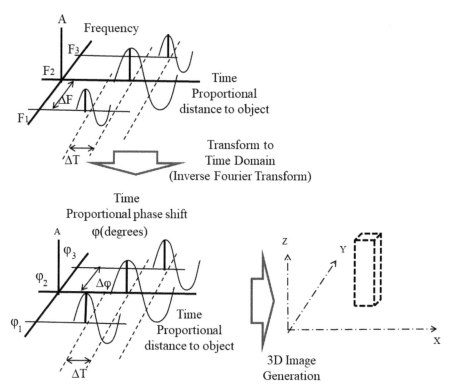

Figure 3.8   Diagram of the system for recording near-field digital holograms.

the reference beam, similar to the principles of optical holography. Scanning a narrow beam antenna around the object surface, as done in some existing imaging radar systems, can be considered as quasi-holograms. However, true digital holograms cannot be recorded by scanning antennas because they do not cover the entire object simultaneously. The loss of signals outside the narrow antenna pattern is not compensated for by scanning the antenna around the object's surface. As a result, the recovered image loses part of the information about the object and suffers from reduced resolution.

Figure 3.8 illustrates the final stage of the holographic radar system, where frequency components are transformed back to the time domain to reconstruct the object in the time–frequency–space domains. It emphasizes the importance of simultaneously covering the entire object or observation area with the antenna array to record true digital holograms and avoid loss of information and resolution.

Figure 3.9 depicts a two-dimensional plane directional antenna array with overlapping monopulse antenna patterns. It illustrates the arrangement

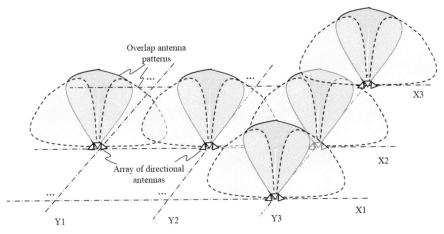

Figure 3.9 Two-dimensional plane directional antenna array with overlapping monopulse antenna patterns. Overlapping of antenna patterns in axis *X* only is shown in the presented picture.

of antenna elements in two axes to cover a wider area of observation and enable the recording of a digital hologram. The overlap of antenna patterns is shown in the *X*-axis in the presented picture.

By positioning multiple antenna arrays in two or more axes, the holographic radar system can achieve a broader coverage area, allowing for the recording of a digital hologram that encompasses the entire observation area. The overlap of the antenna patterns enhances the accuracy and resolution of the system as well as provides additional information about the objects being observed.

While Figure 3.9 displays the overlapping of antenna patterns in axis *X*, the concept can be extended to other axes as well (*Y*, *U*, and *W*) to further improve the image resolution and capture more comprehensive information about the objects. Figure 3.9 showcases a two-dimensional plane directional antenna array with overlapping monopulse antenna patterns, demonstrating how multiple antenna arrays positioned in two or more axes can cover a wider area of observation and enable the recording of a digital hologram.

The proposed technique of recording a digital hologram offers several advantages, including increased image resolution beyond the diffraction limit. The resolution is now primarily limited by the digitizing (sampling) frequency and the stability of the transmitting and sampling frequencies and phases. This is particularly beneficial in applications such as through-wall imaging or medical imaging, where long-wavelength RF electromagnetic signals are used for better penetration through walls or media.

Holographic staring radars, in comparison to scanning radars, operate with a stationary transmit beam and multiple static receiver beams that cover the entire transmit floodlight. The number of receiver beams within the transmit floodlight determines the reception gain and the ability to localize targets in angle. Staring radars offer exceptional information and utility by allowing optimal detection, tracking, and target identification capabilities in an efficient manner. The processing dwell time can be set for specific tasks or multiple parallel processes can be employed with different dwell times. This flexibility is not achievable in mechanically scanned surveillance radars or conventional phased array radars, which have limited time allocated to each sector to maintain complete coverage with acceptable latency.

Holographic radars are particularly well-suited for cluttered environments. They enable target and clutter separation by continuously gathering signals from a large volume of space and fully sampling the amplitude and phase data from every target. Through tracking processes, targets of interest can be effectively separated from clutter. Additionally, the application of an additional reference antenna can enhance opportunities for suppressing scattering medium noise. Holographic radar systems offer distinct advantages in cluttered environments, providing better target and clutter separation, continuous data gathering from a large volume of space, and the ability to fully sample amplitude and phase data from every target. The utilization of a holographic radar system, along with the inclusion of an additional reference antenna, enhances the radar's performance and noise suppression capabilities (Figure 3.10) [9–15].

The near-field RF continuous wave imaging system shown in Figure 3.11 is specifically designed for medical applications. The system consists of several key components:

1. Signal generator: This component generates continuous wave RF signals that are transmitted into the imaging object. The signals cover the entire imaging object, ensuring comprehensive coverage.

2. Transceiver antenna module (TAM): The TAM includes a directional antenna array that is coupled with a multi-channel software-defined radio (SDR). The directional antenna array is responsible for receiving the reflected signals from the imaging object. The SDR enables flexible and programmable processing of the received signals.

3. Multi-channel processor: The received signals from the TAM are processed by a multi-channel processor. The processor performs various signal processing tasks, including digitization, filtering, and reconstruction of the digital hologram.

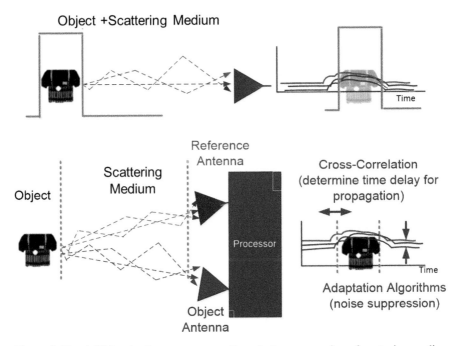

Figure 3.10 Additional reference antenna allows better suppression of scattering medium signals in comparison to the gating of receiving signals.

4. Digital hologram recording: The system employs the concept of digital holography to record the digital hologram of the imaging object. The digital hologram contains valuable information about the shape, size, and dielectric properties of the object. This recorded hologram enables higher resolution imaging compared to traditional methods, achieving resolutions 10–1000 times smaller than the applied wavelength.

The system also utilizes an additional reference antenna, which enhances the imaging resolution and noise suppression capabilities. The reference antenna provides a reference signal that improves the accuracy and precision of the holographic reconstruction process.

The combination of these components enables the near-field RF imaging system to capture high-resolution images of medical objects or tissues. The system's ability to record digital holograms and utilize additional reference antennas contributes to improved imaging resolution, providing valuable information for medical diagnosis and analysis.

Figure 3.12 shows a test setup for recording digital holograms using a monopulse system with four directional antennas and four separate

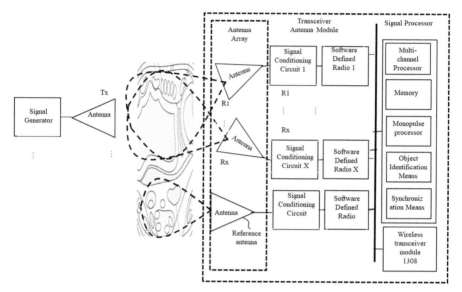

Figure 3.11   Near-field RF imaging radar system for medical applications.

software defined radio (SDR) modules. The components of the test setup include:

1. SDR RTL2832: This is an SDR module with a frequency range of 24–1766 MHz. It is capable of capturing I/Q samples at an 8-bit resolution ratio. The maximum sample rate is 3.2 MS/sec, allowing for high-speed data acquisition.

2. Tuner R820T: The RTL2832 SDR module is equipped with the R820T tuner, which provides features such as a noise figure of 3.5 dB, phase noise of −98 dBc/Hz at 10 KHz, and an image rejection of 65 dBc. These specifications contribute to improved signal quality and performance.

3. Directional antennas: The test setup utilizes four directional antennas, arranged in a monopulse configuration. The directional antennas are designed to capture the reflected RF signals from the imaging object, ensuring comprehensive coverage and accurate recording of the hologram.

The combination of the four separate SDR modules and directional antennas enables simultaneous recording of the digital hologram from multiple perspectives. This approach allows for enhanced image resolution and provides a more comprehensive representation of the object.

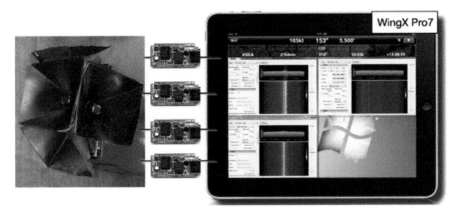

Figure 3.12 Test setup for recording a digital hologram. System comprises the following: SDR RTL2832, frequency range: 24–1766 MHz; resolution ratio: 8 bit with I/Q samples; max samples rate: 3.2 MS/sec; input impedance: 75 Ohm; Tuner R820T: Rafael Micro; noise figure 3.5 dB; phase noise: −98 dBc/Hz @ 10 KHz; max input power: +10 dBm; image rejection: 65 dBc.

The test setup presented in Figure 3.12 serves as a practical implementation of the digital hologram recording system, showcasing the use of SDR technology and directional antennas to capture high-quality holographic data.

3.3 Conclusion

In conclusion, the use of a radar system with multi-beam, multi-axis overlapping antenna patterns and staring capability enables the reception of maximum three-dimensional (3D) information from the entire object or multiple objects within a scattering medium. This approach allows for comprehensive object observation, leading to high-resolution imaging and increased probability of object recognition.

The system's ability to employ multi-axis 3D observation facilitates detailed object representation and improves the accuracy of object recognition. The simultaneous multi-channel signal processing with one-step algorithms and reference signals from overlapping antennas enables high-speed object imaging.

By transforming and processing received signals in the time domain, frequency domain, and multi-axis space domain, the system can further enhance image resolution and facilitate object identification.

Overall, the utilization of a radar system with multi-beam, multi-axis overlapping antenna patterns and staring capability offers significant advantages in terms of imaging quality, object recognition, and processing speed. This technology holds promise for various applications, including medical imaging, through-wall imaging, and radar-based sensing systems.

References

[1] D. Gabor, Microscopy by reconstructed wavefronts, Research Laboratory, British Thomson-Houston Company Ltd., Rugby*, (Communicated by Sir Lawrence Bragg, F.R.S.-Received 23 August 1948-Revised 28 December 1948-Read 17 February 1949).

[2] Dominic Walker, Chief Executive Officer, Aveillant

[3] E. Charles Hendrix, "Radio frequency holography", US patent 3,887, 923 June 3, 1975.

[4] 4. Dean D. Howard, David. C. Cross "Digital radar target imaging with high radar resolution monopulse radar". US patent 3,887,917, June 3, (1975).

[5] Gary Kemp, Holographic radar brings a new dimension to sensing and instrumentation on T&E ranges Collision avoidance, wind farms and scoring, NDIA test and evaluation conference, S4923-P-069 v0.2, March 2011.

[6] S. A. Harman. *A Comparison of staring radars with scanning radars for UAV detection*. Proc. of the 12th European Radar Conference, Paris, France, Sept 2015.

[7] S. A. Harman. *Holographic radar development*. Microwave Journal, Vol. 2, February, 2021, www.microwavejournal.com/articles/35410-holographic-radar-development.

[8] S. E. Lipsky. *Microwave passive direction finder*. SciTech Publishing Inc. Raleigh, NC 27613, 2004.

[9] P. A. Molchanov and A. Gorwara. *Fly Eye radar concept*. IRS2017. International Radar Symposium, Prague, July 2017.

[10] P. A. Molchanov and O. V. Asmolova. *All-digital radar architecture*. Conference: SPIE Security + Defense, DOI: 10.1117/12.2060249, October 2014.

[11] A. Gorwara, P. Molchanov. Multibeam Monopulse Radar for Airborne Sense and Avoid System, Proc. of 2016 SPIE Remote Sensing and Security+Defense Conference, Edinburgh, UK, Paper #9986-3, September 2016.

[12] P. A. Molchanov, M. V. Contarino, "Multi-beam antenna array for protecting GPS receivers from jamming and spoofing signals" US Patent US20140035783, 2014.

[13] P. A. Molchanov. *Tactical radar system for detection of hypersonic missiles and UAS*. US Patent appl. 17/651,800 02/02/2022.

[14] P. A. Molchanov. *Multi-beam multi-band protected communication system*. US patent appl. 17/740,581, 05/10/2022.

[15] P. A. Molchanov. *Passive radar system for detection of low-profile low altitude targets*. US Patent appl. 17/971,582, 10/22/2022.

4

Fast Monopulse Signal Processing

4.1 Monopulse Method of Direction Finding

The monopulse method is a technique used for direction finding in radar systems and other applications. It involves sending additional information in the radar signal to avoid rapid changes in signal strength, making the system more resistant to noise and jamming.

Monopulse radar systems have been in development since the 1960s and are also used in passive systems such as electronic support measures and radio astronomy. The IEEE Standard 145-1983 defines monopulse as a method where direction-finding information is obtainable from a single pulse. It does not specify the number of beams, phase centers, or lobes or any specific antenna topology [1–5].

The monopulse method can be applied to continuous wave, modulated, or pulsed signals. It can be classified into phase, amplitude, or combined monopulse systems based on the type of information used to determine the direction. The monopulse ratio, which contains both amplitude and phase information, is often used for direction finding.

The combination of amplitude and phase monopulse algorithms can lead to the complex monopulse algorithm, which can utilize information from sidelobes by considering the phase shift of signals in relation to the main lobes.

The amplitude comparison method for direction finding is relatively simple to implement and offers good sensitivity and a high probability of signal detection. It typically involves an array of four or more squinted directional antennas to provide 360° coverage.

On the other hand, the phase comparison method for direction finding can provide better bearing accuracy but requires more complex processing. Matching the gains of antennas and their amplifying chains, along with careful design, construction, and calibration, can compensate for hardware limitations.

73

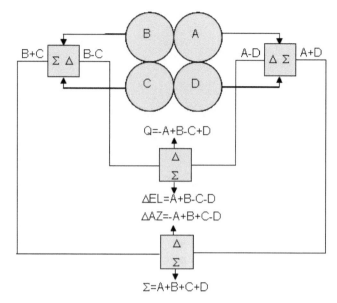

Figure 4.1 Monopulse antenna system and a diagram of monopulse signal processing.

The monopulse method finds applications in radar systems as well as modern communication techniques, offering improved accuracy and robustness in determining the direction of signals.

4.2 Monopulse Antenna Array

The block diagram shown in Figure 4.1 illustrates one of the earlier embodiments of a monopulse antenna system, including the comparator network. The monopulse antenna system is designed to gather direction-finding information with a single pulse, as opposed to emitting multiple narrow-beam pulses in different directions and searching for the maximum return.

The diagram consists of four receiver channels represented by the blue squares, which correspond to the four antenna elements or beams. These receiver channels receive analog signals from the antenna elements and process them using a comparator network.

The signals from the antenna elements are combined and subtracted using 180° hybrid couplers, as indicated in the diagram. This combination and subtraction process allows for the extraction of direction-finding information from the received signals.

By analyzing the resulting signals from the comparator network, the direction of the source of the signal can be determined. This direction-finding information can then be used for various purposes, such as radar tracking, target identification, or communication signal processing [1–5]. The monopulse antenna system and its associated processing provide a more efficient and effective way to gather direction-finding information using a single pulse, improving the accuracy and reliability of the system.

The monopulse uses four antennas (or quadrants of a single antenna). They can be directional antennas as horns, or sections of a flat plate array of radiators, or even sub-arrays of an active electrically scanned antenna (AESA) phased array. The elements are all steered together mechanically (on a gimbal) or electrically (using phase shifters in the case of AESA). The target is illuminated by all four quadrants equally. A comparator network is used to "calculate" four return signals. The sum signal has the same pattern in receive as in transmit: a broad beam with the highest gain at boresight; the sum signal is used to track target distance and perhaps velocity. The elevation difference signal is formed by subtracting the two upper quadrants from the two lower quadrants and is used to calculate the target's position relative to the horizon. The azimuth difference signal is formed by subtracting the left quadrants from the right quadrants and is used to calculate the target's position to the left or right. A fourth signal, called the "Q difference" is the diagonal difference of the quadrants; this signal is often left to rot in a termination, and so the typical monopulse receiver needs only three channels. Sometimes, only a two-channel receiver is used, as the two difference signals are multiplexed into one with a switching arrangement.

Figure 4.2 illustrates the geometry of a target with respect to the elevation angle Θ (theta) in a simple monopulse antenna system with two antenna elements. In this configuration, the target is positioned at a distance L from the upper quadrants of the monopulse antenna. The target is slightly farther away from the lower quadrants, represented by the distance ΔL. The difference in distance ΔL varies as the sine of the angle Θ and the separation between the two antennas. By measuring the difference in received signals between the upper and lower quadrants, the monopulse antenna system can calculate the elevation angle of the target relative to the antenna. This elevation angle information is important for determining the target's position relative to the horizon. The geometry presented in Figure 4.2 is a simplified representation of a monopulse antenna array with two antenna elements. In practice, monopulse systems typically use four antenna elements (quadrants) to provide more accurate direction-finding information in both azimuth and elevation.

The monopulse technique allows for improved accuracy in tracking and localizing targets compared to traditional radar systems, as it provides

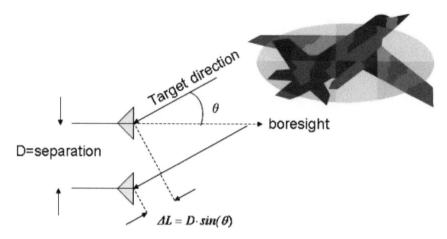

Figure 4.2   Elevation angle for two-antenna-element monopulse antenna array.

simultaneous measurement of both azimuth and elevation angles using a single pulse.

The amplitude and/or phase comparison used in the monopulse method of signal processing is relatively straightforward to implement and offers good sensitivity, high accuracy in direction finding, and a higher probability of signal detection. Typically, an array of four or more squinted directional antennas is used to provide 360° coverage.

To achieve complete coverage, antennas are arranged in a circular array, paired to calculate the signal levels received at each antenna. If there are N antennas in the array, the angular spacing, or squint angle (Φ), is determined by $\Phi = 2\pi/N$ radians (or $360/N$ degrees).

By comparing the signal amplitudes or phases in two adjacent channels of the array, the bearing of an incoming wave front can be obtained. In some cases, three adjacent channels are used to achieve improved accuracy. While it is important to closely match the gains of the antennas and their amplifying chains, careful design, construction, and effective calibration procedures can compensate for any hardware discrepancies.

Monopulse direction finding does not require complex algorithms for direction calculation. Typically, a simple one-step algorithm is used to calculate the ratio of amplitudes or phases in two or more monopulse antennas with overlapping patterns. Each directional antenna is coupled with a separate receiving channel and connected to a signal processor via a digital interface (as shown in Figure 4.3). By digitizing the received signals directly on the antennas, the calculation of direction becomes faster, and the use of

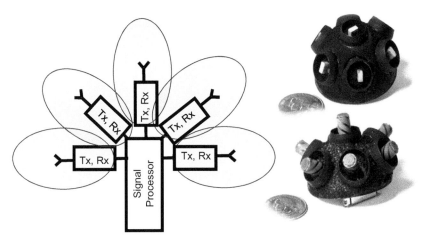

Figure 4.3   Directional antennas are integrating with signal processing for fast direction finding in monopulse antenna array.

reference signals in overlapping antennas can significantly improve directional accuracy, achieving 2–3 orders of magnitude better accuracy.

4.3 Accuracy of Measurement of the Direction

The accuracy of measuring the direction of arrival (DOA) of radiation or signal sources is crucial for determining the pointing accuracy in monopulse systems. DOA estimation involves retrieving the direction information of electromagnetic waves or sources based on the outputs of a sensor array composed of multiple receiving antennas. DOA represents the direction from which a propagating wave arrives at a specific point where the sensor array is located. The measurement of DOA is typically based on the measurement of phase differences between two antennas positioned at a baseline distance in a direction perpendicular to the signal arrival (as shown in Figure 4.4(a)).

In the monopulse method of direction finding (as depicted in Figure 4.4(b)), the use of overlap antenna patterns significantly enhances the accuracy of phase measurements, leading to improved directional accuracy. By employing overlapping antenna patterns, the monopulse system achieves better resolution in measuring phase differences, resulting in more precise determination of the direction of the incoming signal or target reflection. The utilization of overlap antenna patterns in the monopulse method allows for better phase measurement accuracy, which directly translates into improved directional accuracy in determining the source or target's direction of arrival.

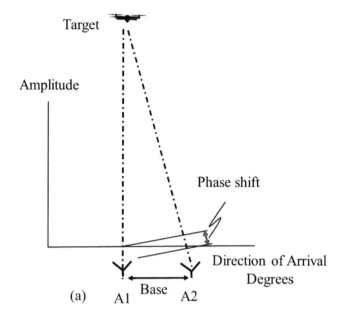

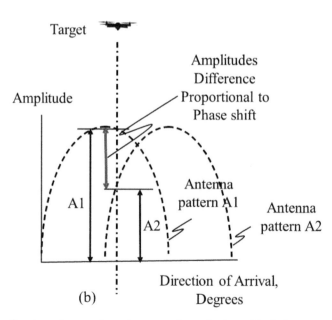

Figure 4.4   Overlap antenna patterns in monopulse method provides better accuracy of phase measurement and corresponding directional accuracy.

In the context of direction finding using the monopulse method, the phase angle difference between antennas, denoted as ψ, is a crucial parameter. This phase difference is compared against the phase front of the arriving signal. The difference in signal path length, represented by the variable S, is determined by the displacement (spatial angular shift) of the antenna aperture, denoted as D. This displacement gives rise to the equation $S = D \sin \varphi$, where φ represents the angle of arrival measured from the boresight and λ represents the wavelength of the signal.

The equation for the phase angle difference ψ is given by [1]

$$\psi = -2\pi \frac{S}{\lambda} = -2\pi \frac{D \sin \varphi}{\lambda} \tag{4.1}$$

where:
φ = the angle of arrival measured from bore sight;
λ = the wavelength.

If A and B are RF voltages measured at the reference boresight and incident antennas, respectively, then

$$A = M \sin (\omega t) \tag{4.2}$$

and

$$B = M \sin (\omega t + \psi) = M \sin (\omega t - \frac{2\pi D}{\lambda} \sin \varphi), \tag{4.3}$$

where M is a common constant defined by signal power. This shows that the angle of arrival φ is contained in the RF argument or phase difference of the two beams for all signals off the boresight axis.

Direction finding based on amplitude comparison methods can provide root mean square (RMS) accuracy smaller than 2° within 100 ns after the direct wave arrives. High-accuracy phase measurements not only enable accurate direction finding but also contribute to high-accuracy imaging.

The space tilt of directional antennas can further enhance the accuracy of phase measurements when the monopulse method of direction finding is applied. Figure 4.5 illustrates the antenna patterns of two spatially tilted directional antennas in a directional antenna array. By introducing spatial tilt, the accuracy of phase measurements can be improved, leading to enhanced direction finding capabilities.

In Figure 4.6, two objects, object 1 and object 2, are depicted along with their corresponding antenna patterns. A small shift of object 1 results in a phase shift, which can be measured by a small change in amplitude. On the other hand, the same shift of object 2 leads to the same phase shift but with

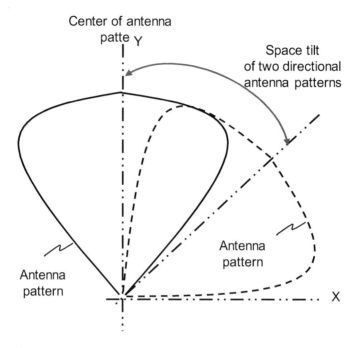

Figure 4.5   Space tilt of two antenna patterns in directional antenna array.

a much larger change in amplitude. This is because object 2 has shifted from the center of the antenna pattern to a slope.

By introducing a space tilt in the two overlap antennas, the phase measurement accuracy can be significantly improved. The ratio of amplitudes in the overlap antennas provides a better indication of the phase shift, allowing for more precise measurement. This enhanced accuracy in phase measurement can be advantageous in the monopulse method of direction finding, as it contributes to improved directional accuracy and overall performance.

High accuracy phase measurements provide high accuracy imaging. If antenna lobes are overlapping or closely spaced, a monopulse antenna array can produce a high degree of pointing accuracy within the beam, adding to the natural accuracy of the space tilted antennas or like in conical scanning system. Classical conical scanning systems generate pointing accuracy on the order of 0.1°, whereas monopulse radars generally improve this by a factor of 10, and advanced tracking radars like the AN/FPS-16 are accurate to 0.006° [10]. High-resolution measurement data allows for the reconstruction of high-resolution images. There are a few accuracy parameters that need to be taken into account to determine approximate reconstructed image

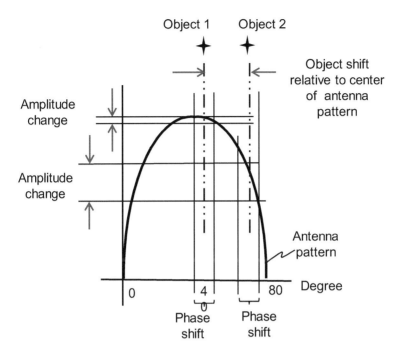

Figure 4.6 Same object shift and corresponding phase shift in antenna leading to much better phase measurement accuracy on antenna pattern slope.

resolution. Range accuracy and range resolution are parameters usually used in radar systems. Range accuracy, δR, is related to range resolution, ΔR, as follows [11]:

$$\delta R \cong \frac{\Delta R}{\sqrt{2SNR}} [m] \text{ for } SNR \gg 1. \tag{4.4}$$

Accurate range measurement is different (though related) from range resolution.

The ability to accurately extract the round-trip time of flight depends on the range resolution, ΔR, and on the signal-to-noise ratio, SNR.

It can be shown that range accuracy, δR, is related to bandwidth, B, and SNR as follows:

$$\delta R \cong \frac{c}{2B\sqrt{2SNR}} [m] \tag{4.5}$$

For example, for SNR >> 1, consider a radar with $B = 2000$ MHz and an SNR of 20 (13 dB). The achievable range resolution, ΔR, is 0.5 m and the achievable range accuracy, δR, is 1 cm.

In general, the uncertainty associated with any measurements is related to the measurement resolution. For measurement with resolution ΔM, the accuracy, δM, is

$$\delta M \cong \frac{\Delta M}{\sqrt{2SNR}}\left[m\right] \text{ for } SNR \gg 1. \tag{4.6}$$

By coherently adding echo signal energy from consecutive pulses, it is possible to effectively increase the illumination energy. This may be thought of as increasing the transmitted power, P_t. The radar transmitter has peak output power and a pulse duration τ, and the transmit pulse energy is: $P_t \cdot \tau$.

Coherently integrating echoes from 10 pulses ($N_{coh} = 10$) produces an SNR equivalent to the case where P_t is 10 times greater.

Alternatively, coherent integration permits a reduction of the transmit pulse power P_t equivalent to the N_{coh} while retaining a constant SNR [12]:

$$\sum_{n=1}^{N_{coh}} \quad \boxed{Tx_n} \quad \boxed{Tx} \quad N_{coh}P_t \tag{4.7}$$

Coherent integration can provide significant signal-to-noise ratio (SNR) improvement by combining multiple coherent pulses to enhance the detection of weak signals. However, it relies on certain assumptions to be satisfied for optimal performance. The first assumption is that the noise in each pulse should be uncorrelated from pulse to pulse. This allows for the coherent summation of the signal while reducing the effects of random noise. If the noise is correlated, it can degrade the performance of coherent integration.

The second assumption relates to the phase variation of the received signal over the integration interval. Ideally, the phase should vary by less than 90° during the integration to maintain the coherency of the signal. If the phase variation exceeds this limit, it can lead to phase cancelation and degradation of the coherent integration.

In cases where the target or object being observed is moving relative to the radar system, the assumption of limited phase variation becomes crucial. If the phase varies significantly over the integration interval due to target motion, additional processing techniques such as range migration and focusing may be required to compensate for the phase variations and maintain coherency.

Regarding monopulse systems, as mentioned, there are three main types: amplitude monopulse, phase monopulse, and combined monopulse systems. These systems utilize the differences or ratios of amplitudes and

Figure 4.7 Monopulse antenna arrays designed for micro-robot applications.

phases of the lobes in a monopulse antenna array to determine the direction of arrival (DOA) of a received signal. They are commonly used for DOA estimation and for nulling undesired signals coming from specific directions.

Normalization operations, such as dividing the difference between lobes by their sum or a reference signal, are often applied to improve the accuracy and robustness of the monopulse measurements. These operations help compensate for variations in the overall signal strength and enable more reliable DOA estimation or nulling of specific directions. Coherent integration and monopulse techniques play important roles in enhancing radar performance, enabling accurate DOA estimation, and mitigating interfering signals. Monopulse antenna arrays designed for micro-robot applications presented in Figure 4.7.

4.4 Conclusion

In conclusion, the monopulse multi-beam array of staring directional antennas offers several advantages for radar and communication systems. The use of multiple directional antennas allows for a 360° full sphere coverage, eliminating the need for scanning and providing continuous illumination of the entire search space. This continuous coverage enables uninterrupted communication with all satellites or observation of multiple targets without time delays.

The use of non-scanning transceiver modules further enhances the system's capabilities by ensuring continuous and simultaneous operation. By employing multiple directional antennas, the system achieves larger signal gain compared to phased arrays, where the signal gain decreases with the number of beams.

The monopulse method, combined with the use of separate receiver chains for each receiving antenna, enables simultaneous and continuous wave or pulse detection with high directional accuracy. The signals received from the directional antennas are processed using one-iteration algorithms, such as ratio of amplitudes, phases, or frequency components. This results in very fast processing times, typically on the order of a few microseconds.

Overall, the monopulse method with multi-beam directional antennas and non-scanning transceiver modules provides efficient and accurate signal detection and processing, making it suitable for a wide range of applications in radar and communication systems.

Non-scanning monopulse system allows a dramatic decrease in transmitting power and at the same time increase in radar range by integrating 2–3 orders more signals than regular scanning radar systems.

Monopulse radar systems can be passive, using ambient RF energy. Monopulse method provides better sensitivity/range and better target resolutions of 2–3 orders than scanning radars.

References

[1] Stephen. E. Lipsky, "Microwave Passive Direction Finder", SciTech Publishing Inc. Raleigh, NC 27613, 2004.

[2] Donald R. Rhodes, Introduction to Monopulse, Mc Graw Hill; New York, Toronto, London, 1959

[3] S.M. Sherman, Monopulse Principles and techniques, Artech House, Dedham, 1984

[4] IEEE Standard Definitions of Terms for Antennas, IEEE Standard 145-1983, The Institute of Electrical and Electronics Engineers (IEEE), 22 June 1983.

[5] Samuel M. Sherman, David K. Barton, Monopulse Principles and Techniques — second edition, ISBN-13: 978-1-60807-174-6, Artech House, Inc., 2011.

[6] Armin W Doerry, Douglas L Bickel, "Two-Channel Monopulse Antenna Null Steering", SANDIA REPORT, SAN D2020-4428, Printed April 2020.

[7] Ashok Gorwara, Pavlo Molchanov, "New high-resolution imaging technology Application of advanced radar technology for medical imaging", Conference: SPIE, Medical Imaging, Orlando, DOI:10.1117/12.2251438, February 2017.

[8] Chris Allen, "Radar Measurements", Course website URL <people.eecs.ku.edu/~callen/725/EECS725.htm>

[9] Radar Set - Type: AN/FPS-16. US Air Force TM-11-487C-1, Volume 1, MIL-HDBK-162A. 15 December (1965).

[10] Wikipedia. Monopulse radar.

[11] A. Gorwara, P. Molchanov, O. Asmolova, "Doppler micro Sense and avoid radar", 9647-6, SPIE, Security+Defense, Toulouse, France, (2015).

[12] Ashok Gorwara, Dr. Pavlo Molchanov, Multibeam Monopulse Radar for Airborne Sense and Avoid System, Planar Monolithics Industries Inc., 7311-F Grove Road, Frederick, MD, USA 21704.

5

Direct Digitizing: Distributed Radar

5.1 Direct Digitizing

Direct digitizing of received signals directly on each directional antenna or antenna array module, synchronized with a common source, offers several advantages in signal processing and imaging. By digitizing the signals at each antenna, it becomes possible to distribute the antennas on a vehicle or around an object, allowing for wave front reconstruction or real-time digital holography in the processor.

The signals received from multiple antennas, each directed toward different predetermined directions, are recorded as a quadrature data matrix or real-time digital hologram. This digital hologram contains real-time information about the amplitude and phase of each signal on each antenna. By utilizing this digital hologram and virtual scanning techniques, it becomes possible to reconstruct the wave fronts of the signals reflected from the object and accurately determine the direction of each point on the object within the covered area. This enables high-accuracy direction finding and imaging for any point of the object.

Direct digitizing and real-time recording of the digital hologram also offer benefits for communication systems. It provides a reliable multi-channel, non-interrupting link budget, allowing for efficient transmission and reception of signals. Additionally, by not relying on scanning beamwidth, the directional accuracy and imaging resolution are not limited by the minimal size of the beamwidth or diffraction limits. Instead, the best directional accuracy is limited by the maximal sampling frequency and stability of the sampling frequency.

Furthermore, with direct digitizing, the image resolution is not dependent on frequency and is not limited by the diffraction limit (Abbe limit). This means that lower frequency electromagnetic waves can be used for enhanced resolution imaging, which is particularly beneficial in medical imaging applications where high-power millimeter wave signals are often used due to their ability to penetrate challenging imaging mediums.

Another advantage of directional digitizing is that the scanning time is not dependent on the beamwidth or the corresponding lowest scanning signal frequency. The time required for virtual (digital) scanning of the entire observation area is determined by the speed of digital processing, which includes factors such as the number of processor channels, the maximal sampling frequency, and the processor speed.

In summary, direct digitizing of received signals on each directional antenna or antenna array module offers numerous benefits, including accurate direction finding, high-resolution imaging, improved communication link budgets, and flexible scanning capabilities. It allows for efficient signal processing and imaging without being limited by beamwidth or diffraction limits, making it a valuable technique in various applications [1–9].

The use of a digital interface in the direct digitizing approach allows for the connection of digitized signals from each antenna to a signal processor. This digital interface serves as a communication link between the antennas and the processor, similar to how neurons connect individual sensors to a fly's brain.

In the case of a fly eye antenna array system, each antenna is equipped with an analog-to-digital converter (ADC) or a software defined radio that includes an ADC. The digitized signals are then transmitted via the digital interface to the signal processor, where they can be processed and analyzed in real time.

The digital interface can be implemented using various technologies such as universal serial bus (USB), microwave or fiber optic waveguides, or even wireless communication. These interfaces ensure a reliable and efficient transfer of the digitized signals from the antennas to the processor.

By directly digitizing the signals at each antenna and using a digital interface, it becomes possible to distribute the antennas (sensors) loosely around the perimeter of an unmanned aerial system (UAS) or a ground vehicle. This flexible distribution allows for optimal coverage and sensing capabilities. The digitized signals can be processed and stored in the processor as a database of real-time wave front information, which can be utilized for various purposes, including holographic object reconstruction if needed.

The sample illustrations in Figures 5.1 and 5.2 demonstrate the distribution of a fly eye antenna array around the perimeter of a UAS and a ground vehicle, respectively. The lower RF range antenna modules in the ground vehicle arrangement are designed as metal constructions to provide additional protection to the vehicle. The use of a digital interface in conjunction with direct digitizing enables the efficient and flexible integration of directional

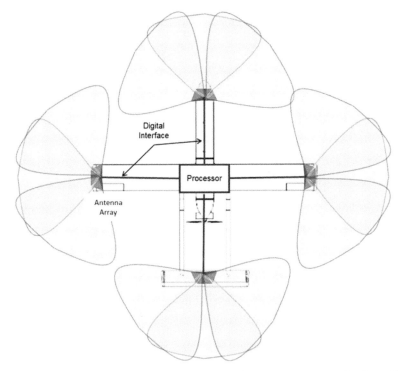

Figure 5.1 Sample of distribution of fly eye or directional antenna array with direct digitizing around the perimeter of UAS.

antenna arrays into various systems, allowing for real-time signal processing, data analysis, and potential holographic reconstruction of objects.

In the direct digitizing approach, the received signals can be directly digitized at each antenna module, enabling the positioning of a fly eye antenna array in an integrated module or the distribution of antenna modules around the perimeter of a UAV. Additionally, the direct digitizing approach facilitates the distribution of antenna modules between a swarm of UAVs.

By using common processing synchronization time signals, the antenna modules can be synchronized and coordinated in their digitizing and data transmission. This synchronization ensures that the digitized signals from each antenna module are aligned and can be effectively processed and analyzed. The use of wireless communication between the swarm of UAVs allows for seamless and efficient data exchange between the antenna modules and the central processing unit. This wireless communication enables real-time data sharing and coordination among the UAVs, enhancing their collective sensing and communication capabilities.

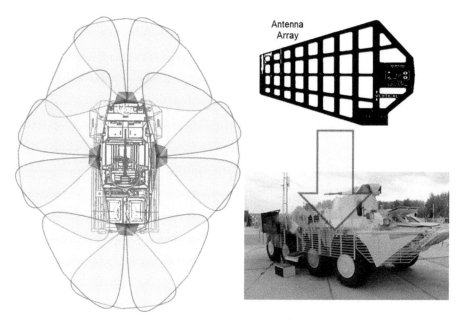

Figure 5.2   Sample of distribution of fly eye antenna array around the ground vehicle. Lower RF range antenna modules are arranged as metal constructions, providing additional vehicle protection.

Figure 5.3 illustrates the concept of direct digitizing and the distribution of fly eye antenna modules either in an integrated module or around the perimeter of a UAV as well as the distribution between a swarm of UAVs. This configuration allows for flexible deployment and coordination of the antenna modules, maximizing the coverage and sensing capabilities of the UAV system. The direct digitizing of received signals, along with the use of common processing synchronization time signals and wireless communication, enables the efficient and distributed positioning of fly eye antenna arrays in UAV systems, enhancing their sensing, communication, and coordination capabilities.

The loose distribution of antennas, not limited by half-wavelength distance between antennas, offers several advantages in antenna array design and functionality. One of the benefits is the ability to arrange multiple antennas or antenna arrays operating at different frequencies within the same aperture. This allows for the integration of multiple functions or sensing capabilities in a single antenna array. For example, in a swarm of UAVs, each UAV can be equipped with a sense and avoid antenna array operating at one frequency while simultaneously incorporating a synthetic aperture antenna radar with

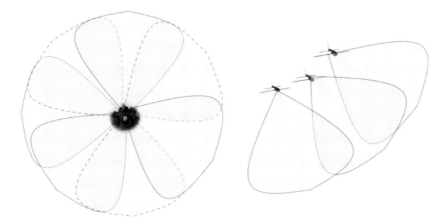

Figure 5.3   Direct digitizing of received signals allows positioning of fly eye antenna array in integrated module or distribute it around the perimeter of UAV or distribution between the swarm of UAVs.

antennas distributed between the UAVs operating at a different frequency. This configuration enables the UAVs to perform both sense and avoid functions and synthetic aperture radar imaging concurrently, utilizing the same swarm antenna array, as depicted in Figure 5.4.

Moreover, the use of frequency-independent traveling wave antennas (TWAs), such as Vivaldi antennas, further expands the design possibilities. TWAs are characterized by their flared structure that provides a gradual transition from a transmission line to free space, resulting in low reflection and broad bandwidth. This allows for the design of ultra-wide bandwidth antenna arrays, capable of operating across a wide frequency range.

The bandwidth of the TWA-based antenna arrays can be significantly larger compared to traditional antenna designs. The gradual transition and inherent impedance matching properties of TWAs contribute to their wideband characteristics. As a result, these antenna arrays can cover a broad frequency range, enabling versatile and multi-functional applications in communication, sensing, and radar systems.

In summary, the loose distribution of antennas not only enables the integration of multiple frequencies and functions within the same antenna array but also allows for the utilization of ultra-wide bandwidth antenna arrays based on frequency-independent traveling wave antennas. These capabilities enhance the functionality and flexibility of swarm systems, enabling diverse sensing, communication, and radar applications [1–9].

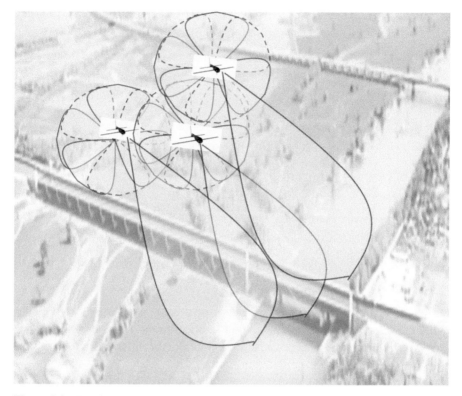

Figure 5.4 Two frequencies and two functions distributed in a swarm antenna array applied as sense and avoid functions in a swarm synthetic aperture radar.

5.2 Direct Sampling Technology

Direct digitizing of signals in antenna array and antenna array systems consists of two important parts. The first part, presented above, is direct digitizing signals of antennas. Direct digitizing combined with digital interface provides possibilities of loose positioning, distributing antennas, or antenna modules (Figures 5.5, 5.6). The second part of direct digitizing in antenna array systems involves direct sampling technology. This technology enables high-speed data conversion between analog and digital signals, allowing for a wider bandwidth of the system. Direct sampling is commonly achieved using analog-to-digital converters (ADCs) and digital-to-analog converters (DACs).

Contemporary ADCs and DACs have advanced capabilities, with some models offering frequencies of up to 20 GHz. This allows for high-speed sampling directly on the antenna or within the software defined radio (SDR) system. Integrating ADCs and DACs directly on the antenna or within the

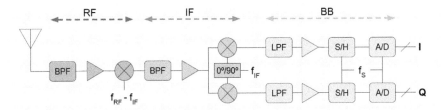

Figure 5.5 Basic architecture of digitizing circuit with mixing (multiplication) performs analog frequency translation and dual-stage mixing allows for relaxed requirements in I/Q demodulation.

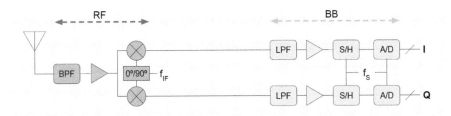

Figure 5.6 Basic direct conversion digitizing circuit with I/Q demodulation at RF. Direct digitizing circuits typically have lower consumption power, smaller SWAP, and can be recommended for monolithic integration.

SDR provides several advantages, including wider instantaneous bandwidths, flexibility, robustness, and improved size, weight, and power (SWAP) characteristics.

The digitizing circuit in direct sampling typically involves mixing or multiplication, which performs analog frequency translation. This process allows for the conversion of signals from one frequency range to another. Additionally, the use of dual-stage mixing provides relaxed requirements in I/Q (in-phase/quadrature) demodulation, which simplifies the signal processing and improves the overall system performance.

Overall, the combination of direct digitizing of signals in antenna arrays and direct sampling technology with high-speed ADCs and DACs offers enhanced bandwidth, flexibility, and robustness in antenna array systems. This enables efficient and reliable data conversion between analog and digital domains, facilitating advanced signal processing and analysis for a wide range of applications [10].

Figure 5.7 shows two samples of schematics for direct digitizing, which allows to increase up to sub-Hz accuracy and apply sub-sampling possibility [10].

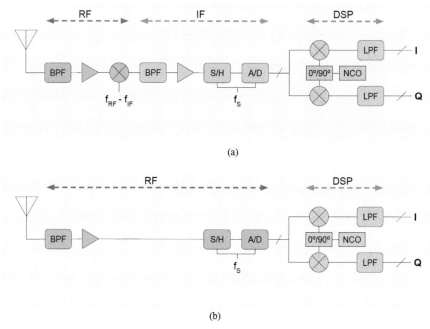

(a)

(b)

Figure 5.7   Direct digitizing circuit with digital I/Q processing offers unlimited precision (sub-Hz accuracy) (a) and with most demanding requirements for S/H and A/D and RF sub-sampling possibility (b).

5.3 Distributed Radar

The development of distributed micro-radar systems represents a significant advancement in the field of passive monopulse direction finding. Stephen E. Lipsky proposed the concept in the 1980s [1]. These systems utilize multiple angularly spaced directional antennas, coupled with front-end circuits, and are individually connected to a direction finder processor through a digital interface. Integrating the antennas with front-end circuits eliminates the need for waveguide lines, which can limit the system's bandwidth and introduce frequency-dependent phase errors.

The concept of passive direction finding systems dates back to the early 20th century. One of the earliest examples is the British Post Office inter-ference finding truck, depicted in Figure 5.8(a). In 1987, the Soviet Union developed the passive radar system known as "Kolchuga," which was later manufactured in Ukraine (see Figure 5.8(b)). The Kolchuga system featured parallel receivers capable of instant discovery and analysis of signals from radio technical equipment (RTE) within the frequency range of 100 MHz to 18 GHz (or sometimes stated as 130 MHz to 18 GHz) [2, 3]. Another passive

(a)   (b)   (c)

Figure 5.8 Passive direction finding systems. (a) British Post Office interference finding truck, 1927. (b) Passive radar "Kolchuga" from Ukraine and (c) passive radar "VERA" from Czech Republic with omnidirectional antennas for a 360° observation.

radiolocator, the VERA system (see Figure 5.8(c)), developed in the Czech Republic, employed time difference of arrival (TDOA) measurements from three or four sites to accurately detect and track airborne emitters.

Additionally, the panorama tracking antenna (Figure 5.9) developed by Aaronia Inc. in Germany offers 3D tracking capability from 9 kHz to 40 GHz with an accuracy of 22.5° (for 4–16 sectors) without the need for rotation. The system boasts super-fast tracking speeds of up to 1 microsecond. However, it is relatively large and may be more suitable for larger aircrafts, weighing approximately 10 kg, and has certain limitations in terms of accuracy [2, 3]. These examples illustrate the evolution of passive direction finding systems and highlight the advancements made in the field. The use of distributed micro-radar systems, combined with advanced signal processing techniques, offers enhanced capabilities for accurate and versatile direction finding across a broad frequency range.

Figure 5.10 illustrates the block diagram of an integrated all-space antenna array architecture with software defined radios (SDRs), direct digitizing, and digital interface to the processor. The architecture consists of several components that enable the reception and processing of additional Doppler-shifted and diffracted spectrum components, which contain information about moving signal sources or objects, including their content or shape.

The transceiver system includes an antenna sub-array, which is responsible for receiving the signals. These signals undergo conditioning and amplification through the front-end circuit (FEC) before being fed into the SDRs. The SDRs perform direct digitizing of the received signals, converting them

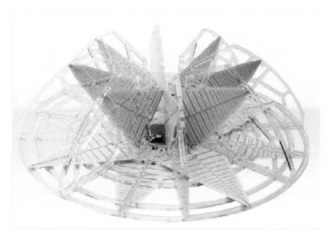

Figure 5.9   3D 360° panorama tracking antenna and screenshot from Aaronia Inc., Germany.

from analog to digital format. The digitized signals are then transmitted to the signal processor via a digital interface.

The signal processor plays a crucial role in processing the received signals. It can comprise an image generator capable of generating one or multi-axis images in various domains, such as time, frequency, space, or a combination of these domains. This allows for comprehensive analysis and visualization of the received signals.

Furthermore, the signal processor includes a monopulse processor, which facilitates simultaneous multi-axis processing of all signals in the receiving chains. This involves calculating ratios of amplitudes and/or phase shifts of signals using a one-iteration adapting algorithm. The goal is to minimize the influence of the surrounding medium on the receiving chain parameters by considering phase shifts in a set of neighboring directional antennas with overlapping antenna patterns. The transceiver system also features synchronization means, which ensure proper coordination and timing among the transmitter chains, receiver chains, monopulse processor, and signal processor. This synchronization is achieved through the digital interface, allowing seamless communication and control between the different components.

The signal processor can be designed to handle simultaneous control of transmitting, receiving, and processing signals on multiple frequencies, multi-frequency signals, or multi-mode signals. This involves incorporating corresponding directional antennas, anti-aliasing circuits, and filtering means in each transmitter and receiver chain to ensure accurate and efficient signal transmission and reception. This integrated all-space antenna array architecture with SDRs, direct digitizing, and digital interface to the processor

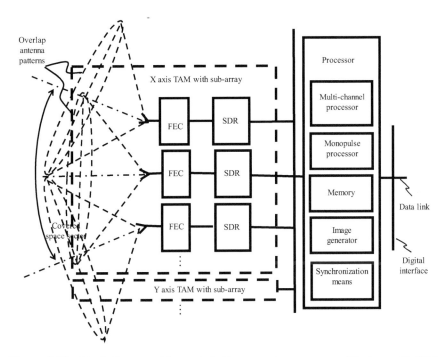

Figure 5.10   Block diagram of integrated all-space antenna array architecture with SDRs, direct digitizing and digital interface to processor. The front-end circuit (FEC) provides conditioning of receiving/transmitting signals.

enables advanced signal processing, multi-axis analysis, and control of signals in various domains, contributing to enhanced performance and versatility in a wide range of applications.

Multi-band, multi-frequency transceivers can be integrated with directional antennas to software-defined modules. Figure 5.11 presents a block diagram of a software-defined direct RF simultaneous sampling transceiver in a single, contained, transceiver antenna module (TAM) connected directly to the antennas [10–14]. The module can also include auxiliary components such as lowpass filters (LPFs), bandpass filters/multi-band (n) pass filters (BPFn), low noise amplifiers (LNAs), high-power amplifiers (HPAs), and circulators. Presented Tx and Rx data conversion components architecture can include transceiver band/service translators (TTs) and clock generators (CGs). Advanced SiP assembly technologies place the FPGA within the same package, allowing to achieve L- through Ka-band module capabilities.

Figures 5.12 and 5.13 illustrate the concept of a UAS (unmanned aerial system) detection radar system using a portable multi-beam staring antenna array integrated with transceiver antenna modules (TAMs). The system is

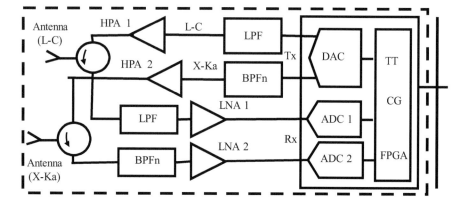

Figure 5.11   Block diagram of software-defined direct RF simultaneous sampling transceiver antenna module.

designed to cover a specific area of observation and can be strategically distributed and camouflaged for discreet operations.

In Figure 5.12, the portable TAMs, which serve as transmitting modules, are distributed throughout the area of observation. These TAMs utilize the multi-beam staring antenna array to detect targets based on their spectrum signature. The received signals from the targets are then processed using monopulse processing techniques to calculate the coordinates of the targets. Each TAM module can transmit information about the detected targets' positions directly to the operator. The operator, equipped with passive modules that receive signals only, can receive the reflected signals from the UAS through the distributed TAMs. This enables the operator to detect and track the UAS from a hidden or remote position.

In Figure 5.13, a similar UAS detection radar system is depicted but with the operator connected wirelessly to the distributed TAMs. The TAMs, still camouflaged and strategically placed, transmit the target position information wirelessly to the operator. This wireless connection allows for remote monitoring and control of the UAS detection process, providing flexibility and convenience for the operator.

These UAS detection radar systems, employing portable multi-beam staring antenna arrays integrated with TAMs, offer a covert and effective means of detecting and tracking UASs within a specific area of observation. The distributed and camouflaged TAMs, combined with monopulse processing and wireless connectivity, provide enhanced situational awareness and target localization capabilities for the operator.

Small-sized inexpensive TAM modules can be installed on ground-based or airborne micro-robots for creating protected communication networks

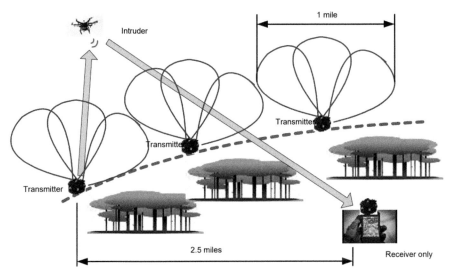

Figure 5.12 UAS detection radar system. Portable, camouflaged transmitting modules distributed in the area of observation. Operators receive reflected UAS signals directly from distributed modules by receiving only passive modules with directional antenna arrays.

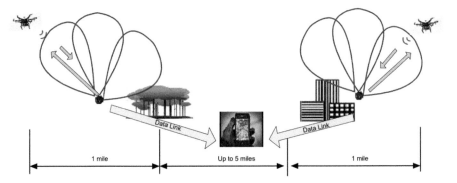

Figure 5.13 UAS detection radar system. Portable, camouflaged transmitting modules distributed in the area of observation. Operators are connected wirelessly with distributed modules.

for surveillance, navigation, and communication and transferring received information to an invisible, not transmitting, operator receiver (Figures 5.14 and 5.15).

5.4 Conclusion

The digitizing of received signals in close proximity to the antennas offers several advantages in the proposed micro-radar system. It allows for a loose

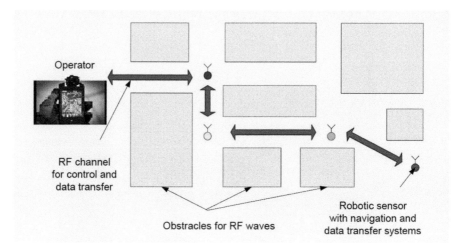

Figure 5.14   Distributed robotic communication network for urban application. Directional communication between micro-robots provides protected communication between micro-robots and allows them to receive data from robotic sensors.

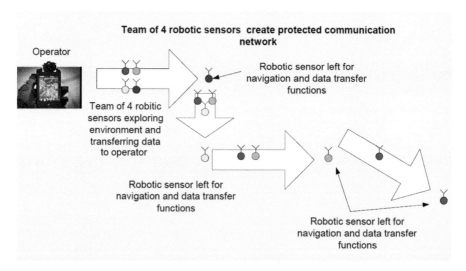

Figure 5.15   Creating a protected communication/navigation network in an area with no navigation signals accessed by a team of four robotic sensors.

distribution of antennas, eliminating the need for waveguides and reducing phase errors associated with them. The accuracy of direction finding in this micro-radar system is primarily determined by the time accuracy of the digital processor and the sampling frequency.

The use of multi-band and multi-functional antennas further enhances the system's capabilities. These antennas can be distributed around the perimeter of a unmanned aircraft system (UAS) or between a swarm/formation of mini/micro UAS. They can be connected to the processor using a digital interface or wirelessly, depending on the specific application. This flexibility enables versatile deployment scenarios and efficient utilization of the antenna resources.

Additionally, the proposed micro-radar system can be used to create a multi-static radar network by distributing expendable micro-radars along the perimeter of a defense object. This network can provide enhanced coverage and detection capabilities, particularly for low-profile, low altitude, and high-speed targets such as small projectiles. The Doppler shift created by these targets can be effectively filtered and detected with a high probability of success.

The micro-radar system can operate in different modes, including passive, monostatic, or bistatic regimes, depending on the specific requirements of the application. Each mode offers unique advantages in terms of detection range, signal processing, and system complexity.

The digitizing of received signals and the use of multi-band, multi-functional antennas in the proposed micro-radar system enable flexible deployment, accurate direction finding, and effective target detection in various operational scenarios.

References

[1] S. E. Lipsky, "Microwave passive direction finder", SciTech Publishing Inc. Raleigh, NC 27613, 2004.

[2] A. Gorwara, P. A. Molchanov, "Distributed micro-radar system for detection and tracking of low-profile, low-altitude targets". Proceedings of the SPIE, Volume 9825, id. 982508 15 pp. May 2016.

[3] P. A. Molchanov, "New Distributed Radar Technology Based on UAV or UGV", Conference, Proceedings of SPIE - The International Society for Optical Engineering · May 2013. https://www.researchgate.net/publication/269325368

[4] P. A. Molchanov, O. V. Asmolova, "All-digital radar architecture", Conference: SPIE Security + Defense, DOI: 10.1117/12.2060249, October 2014.

[5] P. A. Molchanov, M. V. Contarino, "Multi-beam antenna array for protecting GPS receivers from jamming and spoofing signals" US Patent US20140035783, 2014.

[6] P. A. Molchanov "Tactical radar system for detection of hypersonic missiles and UAS", US Patent appl. 17/651,800 02/02/2022.

[7] P. A. Molchanov "Multi-beam multi-band protected communication system" US patent appl. 17/740,581, 05/10/2022.

[8] P. A. Molchanov "Passive radar system for detection of low-profile low altitude targets", US Patent appl. 17/971,582, 10/22/2022.

[9] Mike Kappes, Steven Norsworthy, "Why Direct RF Sampling is a Game Changer", March 2023. IQ-Analog Corporation, 12348 High Bluff Drive, San Diego, California, U.S.A. 92130, visit: www.iqanalog.com

[10] N. Chantier, J. Cochard, "Software-defined direct RF simultaneous sampling multi-band/service transceiver" Teledyne e2v Semiconductors, France, MWJOURNAL.COM, DECEMBER 2022, p. 22–42.

[11] S. Lischi, R. Massini, R. Pilard, D. Stagliano and N. Chantier, "Feasibility study of a fully-digital multi-band SAR system operating at L, C, X and Ku Bands," 7th Workshop on RF and Microwave Systems, Instruments & Sub-systems, 5th Ka-band Workshop, May 2022.

[12] F. Deviere, N. Seller and J. Rohou, "Making history: advanced system in a package technology enable direct RF conversion," White paper, Microwave Journal, Vol. 64, No. 1, January 2021.

[13] T. O'Farrell, R. Singh, Q. Bai, K. L. Ford, R. Langley, M. Beach, E. Arabi, C. Gamlath and K. A. Morris, "Tunable, concurrent multi-band, single chain radio architecture for low energy 5G-Rans," IEEE 85th Vehicular Technology Conference (VTC Spring), June 2017.

[14] K. Ranney, K. Gallagher, D. Galanos, A. Hedden, R. Cutitta, S. Freeman, C. Dietlein, B. Kirk, R. Narayanan, "Software-defined radar: recent experiments and results", Radar Sensor Technology XXII, Proc. of SPIE Vol. 10633, 106331E, 2018.

6

Multi-beam Staring Antenna Array Architecture

6.1 Antenna Array Module

The antenna array module in the proposed system can be designed in different configurations to suit the specific requirements of the application. It can consist of a single array of directional antennas that covers the entire sky or it can be divided into multiple antenna sub-array modules that cover different sub-divided space sectors.

The multi-beam antenna array is constructed using a plurality of antenna elements that form directional antennas with overlapping antenna patterns. These antenna elements can be positioned in a specific order inside a dielectric substrate with a constant or variable dielectric coefficient. Alternatively, they can be arranged as 3D metamaterial substrates, which offer unique electromagnetic properties.

In the case of a multi-beam antenna array, the antenna elements are formed both inside the dielectric substrate and on its surface. This configuration enables the generation of multiple beams with different beamforming patterns, allowing for improved coverage and flexibility in directing the antenna's sensitivity.

Furthermore, directional antennas can also be arranged inside dielectric substrates that are attached to the main transceiver substrate. This approach provides additional flexibility in designing the antenna array and allows for efficient integration of the antenna elements with the overall system. The specific arrangement and configuration of the antenna array module will depend on the desired coverage, beamforming capabilities, and operational requirements of the system [1–5] (Figure 6.1).

In the fly eye system, multiple multi-beam antenna sub-array modules are utilized to provide simultaneous coverage of the entire sky or a designated area of observation. These sub-array modules consist of directional antennas, and the antenna patterns of these directional antennas overlap in

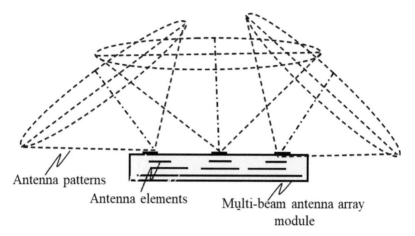

Antenna patterns

Antenna elements   Multi-beam antenna array
module

Figure 6.1   Multi-beam directional antenna sub-array module with antenna elements formed inside and on the substrate surface.

one or more directions. This overlapping configuration allows for the creation of monopulse sub-arrays.

Unlike typical cell phone antenna arrays, where multiple antenna elements are connected to a single transceiver channel, each directional antenna in the proposed system is coupled with a separate transceiver channel. This individual coupling ensures that each directional antenna has its own dedicated transceiver channel and is equipped with a direct digitizer.

The use of separate transceiver channels and direct digitizers for each directional antenna enhances the system's flexibility and capability for accurate signal processing and direction finding. It enables precise beamforming, signal digitization, and independent control of each antenna element, resulting in improved performance and adaptability for a range of applications. The specific configuration and arrangement of the multi-beam antenna sub-array modules may vary depending on the system requirements and operational needs, allowing for customized coverage and directional capabilities (Figure 6.2).

6.2 Multi-beam Transceiver Antenna Array Module

The proposed system utilizes multi-beam antenna modules that integrate multi-channel transceivers with directional antennas, forming software-defined modules. These modules are designed to provide direct RF simultaneous sampling transceiver functionality within a single contained unit. The block diagram of such a module, referred to as a transceiver antenna module

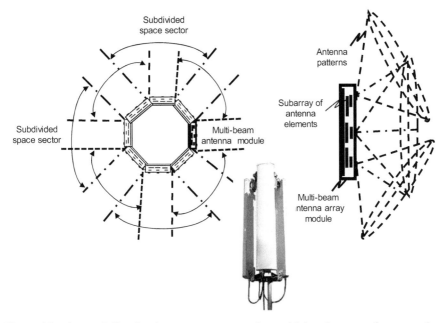

Figure 6.2 Array of directional antennas can comprise multiple sub-arrays of antenna elements. The main difference from cell phone antenna arrays is that each directional antenna is coupled with separate transceiver channel and comprises a direct digitizer.

(TAM), is shown in Figure 6.3 [6–10]. The TAM module can be directly connected to the antennas and is capable of operating in the L- through Ka-band frequency range. It can also incorporate auxiliary components such as low-pass filters (LPFs), bandpass filters (BPFs), multi-band pass filters (BPFn), low-noise amplifiers (LNAs), high-power amplifiers (HPAs), and circulators. These components serve various functions such as signal conditioning, filtering, and amplification.

The architecture of the Tx (transmit) and Rx (receive) data conversion components within the TAM module includes a transceiver band/service translator (TT) and a clock generator (CG). The TT facilitates the translation of signals between different frequency bands or services, allowing for versatile operation. The CG generates the necessary clock signals for synchronization and data conversion processes. Advanced system-in-package (SiP) assembly technologies enable the integration of a field programmable gate array (FPGA) within the same package as the transceiver components. This integration enhances the module's capabilities and allows for efficient processing and control of the transmitted and received signals.

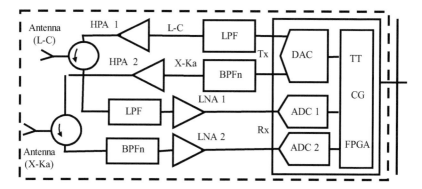

Figure 6.3 Block diagram of software-defined direct RF simultaneous sampling transceiver antenna module.

By combining multi-channel transceivers, directional antennas, and supporting components within a single TAM module, the proposed system achieves a compact and versatile solution for direct RF simultaneous sampling. This architecture enables efficient signal processing, data conversion, and control within a wide frequency range, supporting various applications in the L- through Ka-band spectrum.

6.3 Integration of Transceiver and Antenna Array Module

In the all-digital integral transceiver antenna array module, each directional antenna is coupled with a software-defined radio (SDR) that includes a digitizer, a separate transmitting chain (TC), and a conditioning receiving chain (CRC). The receiving chain includes voltage or current limiters and anti-aliasing circuits to ensure proper signal conditioning. The connection between the directional antennas and the transmitting and receiving chains can be facilitated by switchers (SW) or circulators.

Each SDR is connected to a signal processor and data link through a digital interface, which can be implemented using technologies such as universal serial bus (USB), microwave or fiber optic waveguides, or wireless communication. The signal processor comprises memory for storing executable instructions and implements separate processing of amplitudes, phases, and frequency component shifts in the transmitting and receiving chains. This enables comprehensive signal analysis and manipulation.

The multi-channel processor in the signal processor performs transformations, corrections, and filtering of received signals from the time domain to the frequency domain and space in one or multiple axes. This enhances

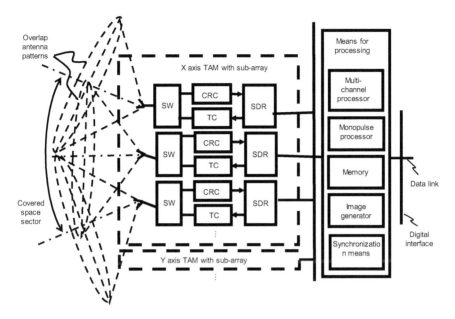

Figure 6.4 Block diagram of integrated all-space antenna array architecture with SDRs, direct digitizing, and digital interface to processor [1–5].

communication reliability, recognition probability, and enables the extraction of a larger number of parameters for object recognition in radar systems.

Additionally, the signal processor includes a monopulse processor that performs simultaneous multi-axis processing of all signals in the receiving chains. It calculates the ratio of amplitudes and/or phase shifts of the signals using a one-iteration adapting algorithm. This processing technique helps minimize the effects of media influence on the receiving chain parameters by considering the phase shifts in a set of neighboring directional antennas with overlapping antenna patterns. Figure 6.4 illustrates the block diagram of the integrated all-space antenna array architecture, highlighting the SDRs, direct digitizing, and digital interface connecting to the signal processor. This architecture enables advanced signal processing, analysis, and image generation in various domains such as time, frequency, space, or combinations thereof.

Direct digitizing of received signals enables the distribution of multiple multi-beam antenna array modules in a strategic manner. These modules can be placed on carriers/satellites, vehicles, or distributed among a swarm or constellation of carriers/satellites to cover the entire sky or a specific area of observation. The transmitter chains, receiver chains, monopulse processor,

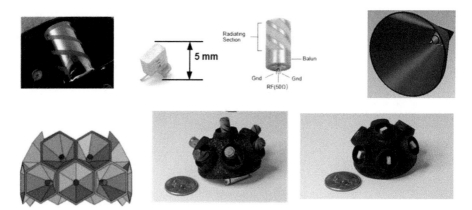

Figure 6.5 Directional antennas applied for designing staring antenna arrays for micro-robots.

and signal processor are connected through synchronization means using a digital interface.

The signal processor is designed to simultaneously control the transmission, reception, and processing of signals on different frequencies. It can handle multi-frequency signals or signals in different modes. Each transmitter and receiver chain is equipped with corresponding directional antennas, anti-aliasing circuits, and filtering means.

The samples of directional antennas used for designing and testing multi-beam staring antenna arrays can have different characteristics. They can operate at a single frequency, multiple frequencies, multiple bands, or even be frequency-independent (such as traveling wave antennas). The direct digitizing of signals on the antennas and antenna arrays is not highly dependent on the distance between the antennas, unlike phased arrays. To reduce interference between closely positioned antennas, screens can be applied.

Figure 6.5 illustrates examples of directional antennas that can be used for designing staring antenna arrays for micro-robots. These antennas play a crucial role in capturing signals and enabling the direct digitization of received signals for further processing and analysis.

6.4 Multi-beam Staring Antenna Array Architecture

The accuracy of a phase antenna array is determined by the size of the beam spot, which is typically minimized at higher frequencies such as centimeter or millimeter wave frequencies. A phase antenna array is constructed with thousands of transmitter/receiver elements that are precisely positioned on a planar substrate, with a half-wavelength distance between antenna elements.

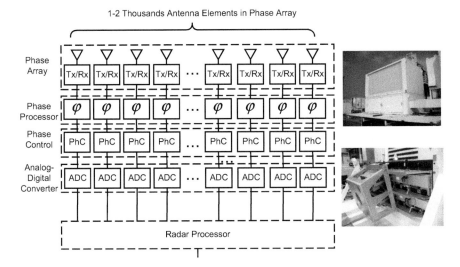

Figure 6.6 The phase antenna array consists of large-sized complicated phase processors (beamformer) and still needs a mechanical scanner if the FOV needs to be more than 120°.

This arrangement allows for precise control of the phase of each element, enabling beamforming and directional beam steering.

However, there are limitations to the coverage area of phase antenna arrays. Even with large-sized and complex phase processors (beamformers), the coverage area is typically limited to around 120° due to phase control errors. To extend the coverage area beyond this limit, mechanical scanning systems are required. These systems involve physically moving the antenna array to steer the beam in different directions. Figure 6.6 illustrates a sample architecture of a regular phase antenna array. The complexity and size of the phase processor (beamformer) are significant, and despite advancements, mechanical scanning systems are still necessary for wider coverage angles. Direct digitizing and software-defined approaches, as mentioned earlier, offer alternative solutions that allow for the distribution of antenna arrays and eliminate the need for mechanical scanning. These approaches provide greater flexibility, precision, and scalability in antenna array systems.

The use of lower frequency antennas in transceiver antenna array modules (TAM) allows for a significant reduction in the number of modules required compared to higher frequency antennas. Lower frequency antennas have a wider area of view (AOV), meaning that each antenna can cover a larger portion of the overall coverage area. For example, with directional antennas that have approximately 120° AOV, it would only take around six antennas to cover a full 360° in azimuth.

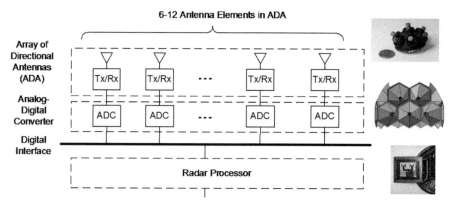

Figure 6.7 Architecture of a fly eye radar. The radar can be all-digital because Tx/Rx coupled with antenna modules can be loosely positioned for small radar size or distributed for small form factor.

By applying reference signals in a set of antennas with overlap antenna patterns, the directional accuracy of the staring antenna arrays can be maintained or even improved. This approach enables simultaneous multi-channel processing, leading to faster target detection and signal processing. Staring antenna arrays require fewer antenna modules and offer faster signal processing compared to traditional phased arrays.

In the case of a multi-beam staring antenna array in a fly eye radar system, multiple directional antennas are coupled with Tx/Rx modules that can be loosely positioned in one point or distributed along the perimeter of a ground vehicle, aircraft, or among robots in a swarm. This flexibility is possible because the front-end modules for receiving and transmitting signals are connected to a central processor through a digital interface. This all-digital radar architecture provides great flexibility in digital modulation and waveform generation.

Figure 6.7 illustrates the architecture of a fly eye radar system, showcasing the loose positioning or distribution of Tx/Rx modules coupled with antenna modules for compact radar size or small form factor applications.

A construction sample of staring antenna array with screening of separate antennas with overlapping in two axes is presented in Figure 6.8.

Passive, receiving only directional antenna module with direct digitizing and direct sampling presented in Figure 6.9.

Samples of two-frequency directional antennas for staring directional antenna array are presented in Figure 6.10. Coupling and direct digitizing of signals in each antenna and monopulse multi-channel processing provide additional multi-frequency multi-function possibilities for these antenna arrays.

Figure 6.8 An array of directional antennas consists of four directional antennas that can be applied for 3D recognition by *X, Y*, and *T* axes or for measurement direction of arrival (DOA).

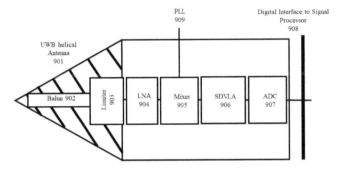

Figure 6.9 Sample of a receiver antenna module design with a wide-band helical antenna.

The additional axis for overlapping antenna patterns provides additional information about target direction, target recognition, or target imaging.

In the context of multi-beam antenna sub-array modules, the directional antennas are designed to cover all-space or the desired area of observation simultaneously. The antenna patterns of these directional antennas overlap in one or more directions, enabling the creation of multi-axis monopulse sub-arrays.

There are different types of antennas that can be used within the sub-array modules. One category includes resonance antennas, such as Yagi antennas and helical antennas, which are designed to operate at specific frequencies. These antennas are arranged to form resonance structures that enhance their performance at those frequencies.

Additionally, frequency-independent traveling-wave antennas can be utilized to extend the bandwidth of the antenna array. These antennas are not limited to a specific resonance frequency and can operate over a wide frequency range. Traveling-wave antennas can be implemented as discrete radiators placed along an axis at specific distances or as continuous radiators that extend along the axis.

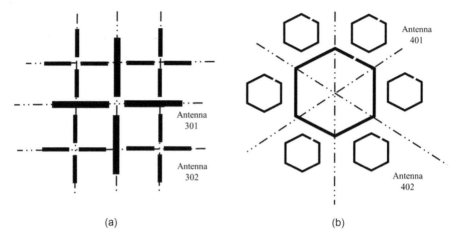

(a)   (b)

Figure 6.10   (a) Two-frequency dipole antenna array with two-axis overlapping antenna patterns, and (b) two-frequency circle antenna array with three-axis overlapping antenna patterns.

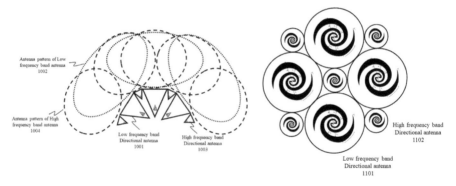

Figure 6.11   Samples of screened two-frequency band antenna arrays.

The direct digitization of signals and the loose positioning of directional antennas enable the combination of antennas within a limited space. This flexibility allows for the efficient use of the available area and the integration of different types of antennas to achieve the desired performance.

Figure 6.11 provides examples of screened antenna arrays operating in two frequency bands, showcasing the versatility and potential configurations of the antenna array design.

Figure 6.12 illustratted transceiver antenna array module with one transmitting channel and a receiving antenna sub-array.

Figure 6.13 illustrates a multi-beam staring antenna array with antenna patterns overlapping in two axes. This configuration enables enhanced

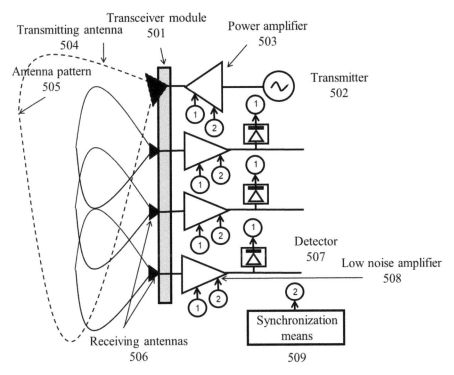

Figure 6.12 Transceiver antenna array module with one transmitting channel and antenna and a receiving antenna sub-array.

accuracy in direction finding. The array incorporates an analog multi-channel signal processor, which facilitates faster non-digital monopulse processing.

The primary objective of this configuration is the detection and tracking of hypersonic missiles and swarms of UAVs. These targets require swift and simultaneous detection across a wide coverage area, followed by instantaneous transfer of the target coordinates to the protection actuator.

By utilizing an analog monopulse processor, the system can detect multiple targets simultaneously in less than 1 millisecond. This is achieved by comparing the ratio of signals received by a set of overlapping antennas in one or two axes. The resulting target coordinates are then swiftly transferred to the actuator, allowing for a prompt and accurate response.

The use of analog processing in this scenario enhances the speed of target detection and tracking, providing real-time information for efficient threat mitigation.

Figure 6.14 presents a design of an array of few sub-array TAMs with different positions of transmitting antennas.

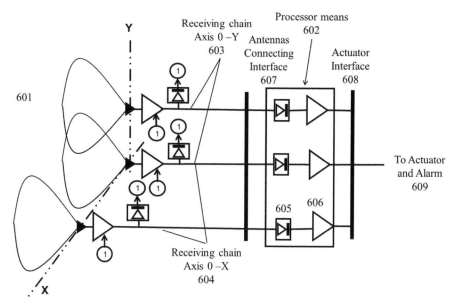

Figure 6.13 Multi-beam staring antenna array wherein antenna patterns overlap in two axes. Analog multi-channel signal processor applied for faster non-digital monopulse processing and transferring target coordinates to the actuator.

6.5 Conclusion

In conclusion, the staring beams directional antenna array presented in this context offers several advantages and capabilities. It addresses the challenge of providing comprehensive coverage and simultaneous multi-channel, multi-band, and multi-function operations with continuous illumination of multiple objects. By utilizing overlap antennas and multi-axis distribution, the array can capture maximum information from multiple sources or objects.

The coupling of each directional antenna with separate transmitting and receiving channels, along with direct digitizing, enables the reception of a digital hologram that describes the entire 3D object simultaneously and at a faster rate compared to scanning systems. Monopulse processing and the use of reference signals from overlapping antennas contribute to higher direction accuracy and improved clutter/noise and media influence suppression.

The system's automatic gain control circuit allows for a separate control of transmitting power and receiver gain in each subdivided sector, making it suitable for use in urban and mountainous areas. Furthermore, the distribution of directional antennas enhances system resilience, as each antenna

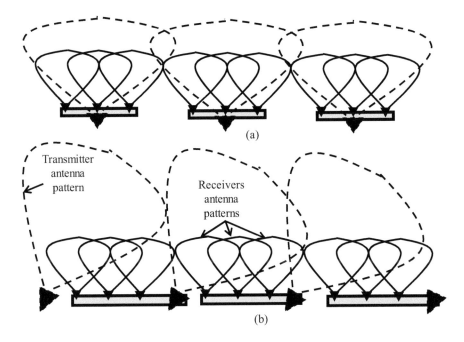

Figure 6.14 Distributed sub-array TAM design with different positions of transmitting antennas.

covers a specific subdivided sector and is less vulnerable to external electromagnetic pulses.

The simultaneous correlation and integration of uninterrupted signals from every point in the observation area enable the detection of low-level signals and aid in signal recognition and classification through the use of multi-axis, multi-frequency diversity signals, polarization modulation, and intelligent processing.

Direct digitizing and synchronization of receiving signals directly on the directional antennas through microwave or optical means allow for loose distribution of antennas without the need for complex phase adjustment matrices.

The directional antenna array can be designed to be multi-band, taking advantage of the frequency-independent nature of the antenna spacing and eliminating the need for a beamforming module. Additionally, the system is compact, lightweight, and can be portable or mounted on light vehicles or small drones due to its small size and weight. Overall, the proposed antenna system offers significant advantages in terms of coverage, accuracy, versatility, and portability.

References

[1] P. A. Molchanov, M. V. Contarino, "Multi-beam antenna array for protecting GPS receivers from jamming and spoofing signals" US Patent US20140035783, 2014.

[2] P. Molchanov "Multi-beam multi-band antenna array module" US patent appl. 17/971,616, 10/23/2022.

[3] P. Molchanov "Tactical radar system for detection of hypersonic missiles and UAS", US patent appl. 17/651,800, 02/02/2022.

[4] P. Molchanov "Passive radar system for detection of low profile low altitude targets", US patent appl. 17/971,582, 10/22/2022.

[5] P. A. Molchanov "Multi-beam multi-band protected communication system" US Patent appl. 17/740,581, 05/10/2022.

[6] N. Chantier, J. Cochard, "Software-defined direct RF simultaneous sampling multi-band/service transceiver" Teledyne e2v Semiconductors, France, MWJOURNAL.COM, DECEMBER 2022, p. 22–42.

[7] S. Lischi, R. Massini, R. Pilard, D. Stagliano and N. Chantier, "Feasibility study of a fully-digital multi-band SAR system operating at L, C, X and Ku Bands," 7th Workshop on RF and Microwave Systems, Instruments & Sub-systems, 5th Ka-band Workshop, May 2022.

[8] F. Deviere, N. Seller and J. Rohou, "Making history: advanced system in a package technology enable direct RF conversion," White paper, Microwave Journal, Vol. 64, No. 1, January 2021.

[9] T. O'Farrell, R. Singh, Q. Bai, K. L. Ford, R. Langley, M. Beach, E. Arabi, C. Gamlath and K. A. Morris, "Tunable, concurrent multi-band, single chain radio architecture for low energy 5G-Rans," IEEE 85th Vehicular Technology Conference (VTC Spring), June 2017.

[10] K. Ranney, K. Gallagher, D. Galanos, A. Hedden, R. Cutitta, S. Freeman, C. Dietlein, B. Kirk, R. Narayanan, "Software-defined radar: recent experiments and results", Radar Sensor Technology XXII, Proc. of SPIE Vol. 10633, 106331E, 2018.

7

Measurement Techniques

7.1 Measurement of Radar Element Parameters

Figure 7.1 shows the physical arrangement of the two directional antennas, along with the setup for the measurement. The setup includes a digital signal generator (ADF435X EVAL BD) and an oscilloscope. The digital signal generator is used to generate the test signals, while the oscilloscope is used to capture the signals and analyze their phase and shape. The oscilloscope screen pictures demonstrate the lock loop oscillator control, which ensures accurate frequency synchronization, and provide visual representations of the measured signals' phase and shape. These measurements help evaluate the performance and characteristics of the directional antennas and their overlap antenna patterns. This measurement technique allows for the assessment and analysis of the multi-beam staring antenna array's behavior and performance, particularly in terms of the antenna patterns and their angular shifts.

Figure 7.2 presents the antenna patterns of an array with two directional antennas that have a 90° angular tilt or shift, as compared to a theoretical model. The graph shows the measured antenna patterns at a frequency of 978 MHz and compares them with the expected patterns based on the model. The maximum amplitude error, relative to the model, is approximately 2–3 dB. This means that the measured patterns deviate from the theoretical model by a small amount in terms of amplitude. While this error is not smaller than what would typically be observed in separate regular directional antennas, the use of a reference beam in a monopulse antenna array allows for the suppression of parasitic reflections, noise, and the influence of the transferring medium.

Monopulse antenna arrays have the potential to significantly improve directional accuracy compared to regular directional antennas. In fact, tracking radars such as the AN/FPS-16 can achieve accuracy as precise as 0.006 degrees. This level of accuracy is achieved through the use of monopulse processing, which compares the ratio of signals received in a set of overlapping antennas.

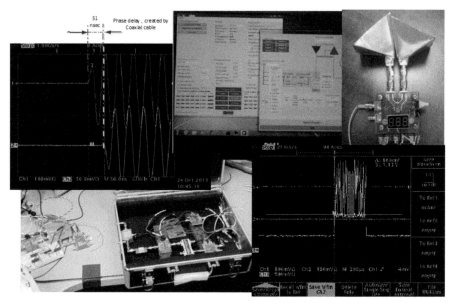

Figure 7.1   Setup for testing a multi-beam staring antenna array that consists of two angularly shifted directional antennas. The measurement technique involves transmitting signals through separate antennas and observing the overlapping antenna patterns. The frequency used for the measurement is 978 MHz.

In Figure 7.3, the test results of a transmitting element with a dielectric resonator are shown, specifically focusing on the frequency characteristics. The dielectric resonator is employed to enhance the frequency stability of the generated signals. The graph illustrates the measured frequency response of the transmitting element. It provides information on how the element performs across different frequencies, typically in the operating range of the antenna array. The stability of the frequency is a critical factor in ensuring accurate and reliable communication and signal processing in staring antenna arrays. By utilizing a dielectric resonator, which is designed to have specific resonant frequencies, the transmitting element can achieve improved frequency stability. This stability helps to maintain the desired operating frequency and minimize any frequency variations that could potentially affect the performance of the antenna array.

The measurement results presented in Figure 7.3 provide an assessment of the frequency characteristics of the transmitting element and demonstrate the effectiveness of incorporating a dielectric resonator for enhancing frequency stability in staring antenna arrays.

In Figure 7.4, a direction of arrival (DOA) detector specifically designed and tested for fly eye radar application is shown. The DOA detector is used

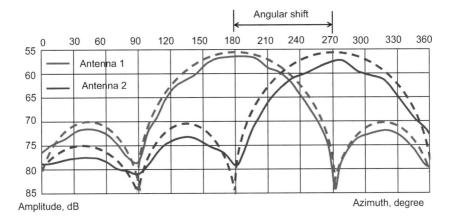

Figure 7.2 Result of the test of the antenna patterns of two directional antennas with an angular shift of 90° compared to the theoretical model. Measuring is made on a frequency of 978 MHz.

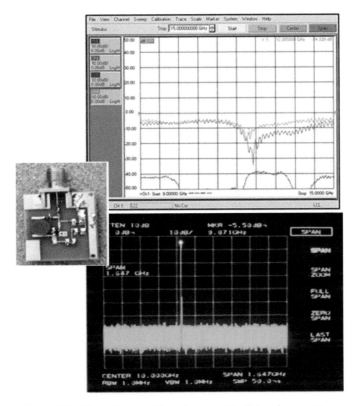

Figure 7.3 Test of transmitting element with dielectric resonator.

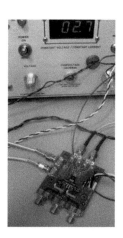

Figure 7.4 DOA detector designed and tested for the estimation of the direction accuracy for a two-antenna array with overlapping antenna patterns. Estimation of bearing accuracy.

to estimate the direction accuracy for a two-antenna array with overlapping antenna patterns.

The purpose of the DOA detector is to determine the direction from which a signal is arriving at the antenna array. This information is crucial for accurately locating and tracking targets in radar systems. By analyzing the phase and amplitude differences between the signals received by the two antennas, the DOA detector can estimate the angle of arrival of the signal.

The graph in Figure 7.4 represents the estimation of the bearing accuracy achieved by the DOA detector based on phase and amplitude measurement for the two-antenna array. It demonstrates that the direction accuracy of the two-antenna array, as determined by the DOA detector, is at least 10 times better than the direct phase measurement.

This improvement in direction accuracy is attributed to the use of overlapping antenna patterns in the array, which allows for more precise estimation of the signal's direction. By comparing the signals received by the antennas and analyzing their phase and amplitude differences, the DOA detector can provide a more accurate estimate of the signal's angle of arrival.

The DOA detector, as depicted in Figure 7.4, showcases the successful design and testing of a system capable of achieving high direction accuracy for a two-antenna array with overlapping antenna patterns in the context of fly eye radar applications.

The fly eye radar range and the accuracy of range measurement have been carried out by phase measurement using the monopulse method. The

Phase modulated radar pulse signal

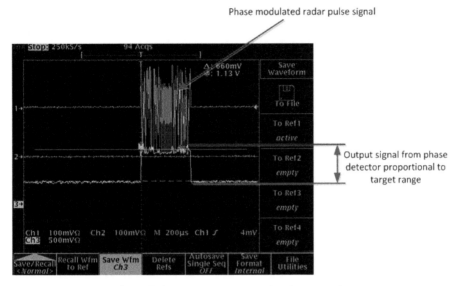

Output signal from phase detector proportional to target range

Figure 7.5   Fly eye radar range bench test result.

output signal from the phase detector is shown in the screenshot presented in Figure 7.5.

7.2 Hardware Component Performance

In general, multi-beam directional antenna arrays are designed to provide coverage over a wide area with multiple beams simultaneously. The antennas may have different designs and configurations based on the specific application, frequency, bandwidth, gain, and directivity requirements. Different companies may have their own designs and technologies for multi-beam antenna arrays. Directional antennas are the main components of multi-beam staring antenna array. An all-space antenna array does not require design directional antennas with some special parameters. Figure 7.6 presents multi-beam antenna arrays designed by different companies for different applications.

Antennas with small size, high Q, and good enough gain have high reactive fields with rapid spatial variation and can be separated in small enough volume. Small-sized antennas with high Q have better separation possibility in space than large antennas with smaller Q. Directional antennas have better separation possibility than omnidirectional antennas. The application of low loss materials and the arrangement of antennas can provide good enough space separation for few small helical directional antennas.

122 *Measurement Techniques*

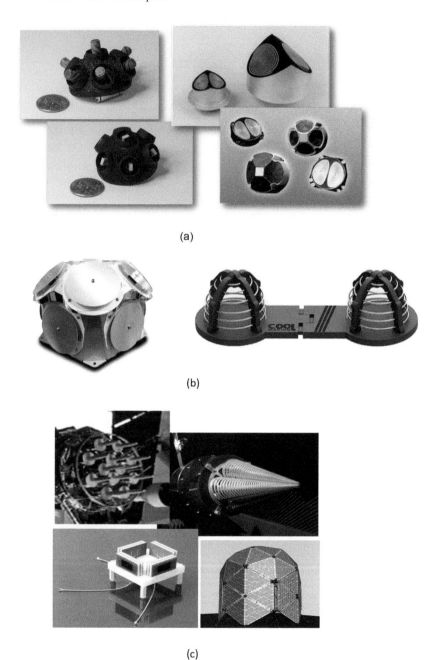

(a)

(b)

(c)

Figure 7.6 Multi-beam directional antenna arrays designed by AMPAC Science Inc. and Cobham Inc. (a), Micro-ant.com Airidium Centus HGA-2 and DJI Mavic 2 Pro (b), and other directional antenna arrays constructions (c).

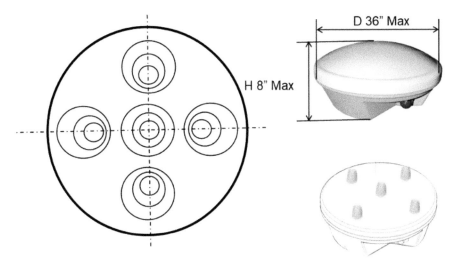

Figure 7.7 Wide bandwidth dielectric directional antenna array designed by IPD Scientific LLC for aircraft applications. The array consists of five tilted active dielectric directional antennas to form a monopulse antenna array.

Few small directional antennas isolated in space can provide incredible protection possibilities against jam, spoof, and undesirable access to GPS, communication, or data link.

A wide bandwidth antenna array designed for aircraft application is presented in Figure 7.7. Dielectric antennas provide focusing of beams, increasing directivity and wide bandwidth. Traveling-wave dielectric antennas can pack to limited size space and be applied in low-frequency RF range.

Testing of interference signals of four transceiver antenna modules (TAM) installed on one drone platform has been done using the setup presented in Figure 7.8. Outputs of the four TAM were connected to LEDs for visualization of the detected signals.

Directional antenna arrays were designed for UAV applications (Figures 7.9 and 7.10). The processing of Doppler signals by a signal processor allows to apply it for the detection of objects consisting of metal and dielectric materials.

Directional antennas arranged in different directions can provide additional protection for an all-space system against jammers, spoofers, or electromagnetic pulses (EMPs). By distributing directional antennas throughout the system, it becomes possible to determine the direction of arrival for signal sources, which can help identify the presence and location of jammers or spoofers.

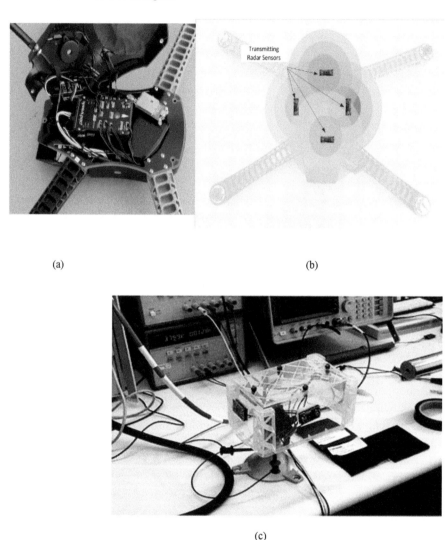

(a)

(b)

(c)

Figure 7.8 Distribution of TAM around drone perimeter (a). Test diagram for measurement interference between four modules distributed around drone platform (b) and test setup (c) [1–3].

This all-digital antenna array architecture with distributed directional antennas enables the system to obtain a three-dimensional space vector to the jammer. By verifying the direction of arrival for signals from various sources, the system can effectively protect against jamming and spoofing attempts. This protection extends not only to radar systems but also to communication systems and any navigation constellation system, regardless of whether the

Figure 7.9 Testing of radar system comprising four transceiver antenna modules distributed with digital interface around the perimeter of a Solo drone [4, 5].

signals are encrypted or unencrypted. The distributed nature of the directional antennas allows for protection against a potentially unlimited number of jammers, even if they are closely positioned. This enhanced protection capability is crucial for ensuring the integrity and reliability of the system's operations in the presence of intentional interference.

References

[1] Pavlo A. Molchanov, A. Gorwara, "Fly Eye radar concept". IRS2017. International Radar Symposium, Prague, July 2017.

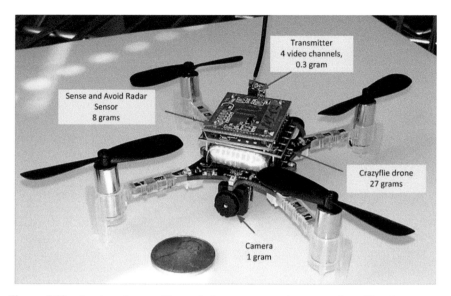

Figure 7.10   A micro-drone with a wireless camera and sense and avoid radar sensors designed and tested for indoor applications.

[2] P. Molchanov, A. Gorwara Fly Eye Radar. Detection Through High-Scattered Area. SPIE Defense+Security Conference, April, 2017

[3] A. Gorwara, P. Molchanov. Multibeam Monopulse Radar for Airborne Sense and Avoid System, *Proc. of 2016 SPIE Remote Sensing and Security+Defense Conference,* Edinburgh, UK, Paper #9986-3, September 2016.

[4] A. Gorwara, P. Molchanov, Distributed Micro-Radar System for Detection and Tracking of Low Profile, Low-altitude Targets, *Proc. of 2016 SPIE Security+Defense conference,* Baltimore, MD, USA, Paper #9825-7, April 2016.

[5] P. Molchanov All Digital Radar Architecture. Paper 9248-11, Security+Defense Conference, Amsterdam, September 25, 2014

8

Fast Radar for the Detection of Hypersonic Missiles and UASs

8.1 Explanation of the Problem

The concepts of a hypersonic and supersonic weapon that could be controlled and maneuvered have been studied for decades, and translating them into development and production has only now become feasible, thanks to recent technological advances. Count hypersonic weapons among its military capabilities can anticipate a number of crucial advantages, all related to an increased degree of uncertainty posed by these weapons to an adversary in the event of a conflict. The velocity with which hypersonic weapons would be able to reach their targets reduces the adversary's ability to either relocate or respond before the strike occurs. The weapons' maneuverability allows them to travel on unpredictable trajectories, making it difficult to track and destroy them before they successfully penetrate advanced air and ballistic missile defenses. Lower signatures and an ability to fly at lower altitudes also compound the challenge of finding, targeting, and intercepting hypersonic vehicles for current missile defense systems [1].

The problem outlined in the given information revolves around the advancement and development of hypersonic weapons by countries, which has raised concerns. These countries have conducted numerous tests and invested significant funding in hypersonic weapons research and development, drawing attention to their capabilities and advancements in this area.

Hypersonic weapons offer several advantages that create uncertainties for adversaries in the event of a conflict. The high velocity at which these weapons travel significantly reduces the time available for the adversary to react or relocate before being struck. The feasibility of developing and producing hypersonic weapons has increased in recent years due to technological advancements. While the concept of hypersonic and supersonic weapons has been studied for decades, the recent progress in technology has made their practical implementation possible.

Figure 8.1   Russian designed hypersonic Zircon missile.

Figure 8.1 depicting Russia's hypersonic Zircon missile and Figure 8.2 showcasing China's demonstrated hypersonic missile serve as visual representations of the technological advancements achieved by these countries in the field of hypersonic weapons.

The emergence of hypersonic technologies and weapons has introduced a significant shift in the military landscape. The high velocities and maneuverability of hypersonic weapons offer distinct advantages that differentiate them from traditional missile capabilities. However, these advancements also pose challenges for existing systems and infrastructure. The speed at which hypersonic weapons travel presents a time-sensitive scenario, where traditional computing systems may struggle to compute fire control solutions, establish communication with command authorities, and complete engagements in a timely manner. The rapid decision-making required to counter hypersonic threats demands more efficient and advanced software-based tools for real-time threat visualization and decision-making.

Furthermore, the existing capabilities of combatant commands to process the large amounts of collected radar, flight test, and shared intelligence data may be inadequate in dealing with the complexities of hypersonic weapons. This highlights the need for improved data processing and analysis capabilities to effectively understand and respond to hypersonic threats.

Protection against the tactical use of hypersonic missiles and maneuvering glide vehicles becomes a critical concern for various defense assets

Figure 8.2 China demonstrated hypersonic missile. Sound speed 767 mph and advertised hypersonic missiles speed 6000 mph, corresponding to Mach 5 or 268 km per second.

such as surface ships and air bases. Ensuring the defense and security of these assets in the face of hypersonic threats requires continuous vigilance and a focus on developing effective defensive measures. The rise of hypersonic technologies necessitates attention and investment in areas such as software-based decision-making tools, improved data processing capabilities, and tactical defense strategies to address the challenges posed by hypersonic weapons.

8.2 Existing Missile Defense Systems

Sectored air and missile defense radars:

The traditional sectored air and missile defense radar systems are designed to detect ballistic missiles within a predetermined sector of space, providing effective coverage for known threats. However, in today's evolving threat landscape, where unmanned aerial vehicles, cruise missiles, and hypersonic

Figure 8.3   Sectored air and missile defense radar.

boost glide vehicles pose significant challenges, threats can emerge from various directions.

The sector-based approach, although suitable for past ballistic missile threats, may not be sufficient to address the complexities of modern multi-domain threats. To effectively defend against such threats, an air and missile defense sensor architecture is needed that can monitor and scan in all directions simultaneously.

By adopting an all-directional approach, defense systems can enhance their situational awareness and improve response capabilities. This includes the ability to detect and track threats from any direction, including unconventional attack vectors. It is crucial to adapt to the changing nature of threats and ensure comprehensive coverage to counter emerging air and missile threats effectively. Figure 8.3 illustrates the concept of sectored air and missile defense radar, highlighting the need for a broader, all-directional sensor architecture to address the evolving threat landscape.

Protection against omnidirectional threats:

The evolving aerial-threat environment, particularly after the Cold War, has become more complex and contested. Traditional air and missile defense

Figure 8.4 Radars with mechanical moving antenna array (Photo Credit: 35th Air Defense Artillery Brigade Public Affairs, Missile Defense Agency).

systems were primarily designed to counter threats with relatively predictable trajectories. However, the current threat spectrum has become far more diverse and challenging.

In today's environment, even adversaries with limited resources can deploy drones and cruise missiles that are capable of maneuvering around sectored sensors, exploiting their blind spots and posing a significant threat. Additionally, the presence of maneuvering ballistic, supersonic, and hypersonic missiles in the upper atmosphere further complicates the detection, tracking, and engagement of these threats.

Russia has demonstrated the effectiveness of integrating electronic attack capabilities, drones, artillery, and short-range ballistic missiles, showcasing the need for comprehensive defense measures (Figure 8.4). The sectored ground-based radars depicted in the image utilize mechanical moving antenna arrays with limited azimuth coverage (approximately 120° and 180°, respectively).

To effectively counter these omnidirectional threats, it is essential to adopt advanced sensor architectures and technologies. AESA wide blind areas on top and bottom cannot provide protection against omnidirectional threats Figure 8.5. This includes the implementation of all-directional radar systems, improved electronic attack countermeasures, and integrated defense systems that can detect, track, and engage threats from any direction. By enhancing situational awareness and developing robust defensive capabilities, it will become possible to mitigate the risks posed by the diverse and complex threat environment of today.

There is need to take radars to 360°. Some are already 360-facing, which means that you can see threats coming from multiple different locations. This reduces chances of targets or incoming missiles flying around your radars. so 360° is very important to defeating next generation threats [2].

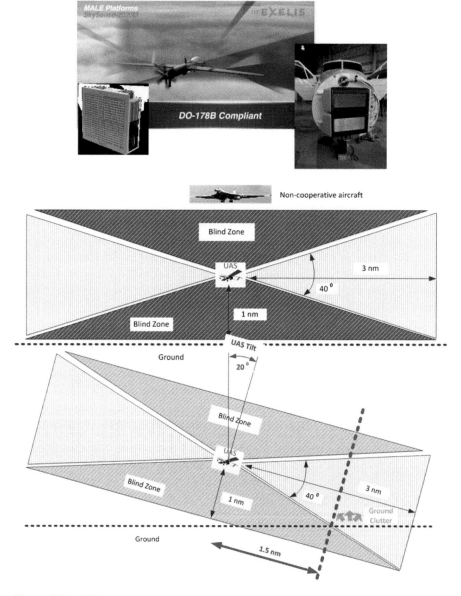

Figure 8.5   AESA radar systems are installed in manned or unmanned airplanes and can provide detection and tracking of multiple targets in forward looking areas. Wide blind areas on top and bottom cannot provide protection against omnidirectional threats. Small tilt of airplane/radar can blind radar by ground clutter.

8.3 Fast Tactical Radar for the Detection of Hypersonic Missiles and UAVs

The requirements for an ideal tactical radar antenna array, specifically designed for detecting and tracking multiple targets simultaneously, include:

1. Multi-beam coverage: The antenna array should be capable of providing coverage for the entire sky, enabling detection and tracking of multiple targets without the need for scanning or switching beams. This ensures that no time is lost in scanning and allows for continuous monitoring of all targets.

2. Continuous tracking: The antenna array must have the ability to continuously track each of the multiple targets, providing maximum information about their movements and trajectories.

3. Simultaneous signal processing: The radar system should be capable of processing signals from multiple targets simultaneously. It should have multi-channel, multi-frequency, and multi-functional capabilities, allowing for comprehensive analysis and detection of various target characteristics.

Traditional phased array antennas with scanning beams do not meet these requirements as the short illuminating time for each target and the need for beam switching limit the information received. Additionally, reducing the scanning beam for better accuracy leads to increased detection time. Moreover, phased array systems require large beam forming means, increasing their size and weight. MIMO (multiple-input multiple-output) antenna arrays are also not suitable as they cannot track targets from multiple directions simultaneously and require switching between beams, resulting in a loss of detection time.

To address these requirements, a tactical radar system based on a fly eye antenna array is proposed. This system employs a multi-beam staring directional antenna array that provides 360° coverage without the need for scanning or interrupting signals (Figure 8.6). It utilizes the monopulse method, either with continuous wave or pulse signals, for fast and accurate detection of hypersonic missiles and UAVs. The system can be designed for multi-frequency and multi-mode operation, including communication, control, and navigation applications. The use of automatic gain control in each beam enables the system to be effective in urban or mountainous areas. The antenna array can be distributed around a vehicle or connected to a network for enhanced coverage and protection.

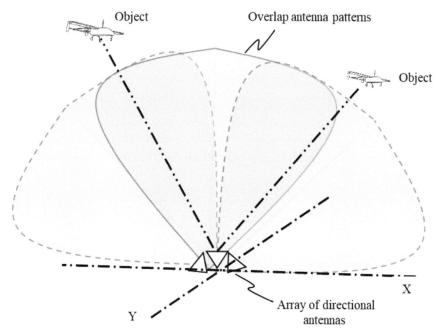

Figure 8.6   All-space antenna array for fast detection and recognition of hypersonic missiles and UAV.

The monopulse method, as described by S. E. Lipsky in US Pat. No 4257047, 1981, utilizes a plurality of fixed, narrow beamwidth antennas to achieve omnidirectional coverage. It involves the use of two pairs of directional antennas, positioned by azimuth and elevation boresight, for microwave direction finding. The method considers three categories of angle sensing: amplitude, phase, or a combination of both ($\Sigma - \Delta$).

The phase angle difference, as measured in each antenna, compared against the arriving signal phase front, is denoted as ψ. The difference in signal path length is defined by the equation, $S = D \sin \varphi$, which depends on the antenna aperture displacement (spatial angular shift) D. Letting φ be the phase lag caused by the difference in the time of arrival between two signals gives:

$$\psi = -2\pi \frac{S}{\lambda} = -2\pi \frac{D \sin \varphi}{\lambda} \qquad (8.1)$$

where φ is the angle of arrival measured from boresight and λ is the wavelength.

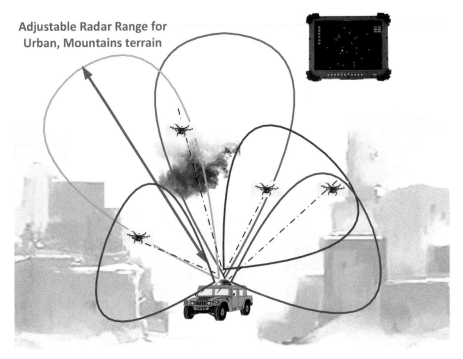

Figure 8.7   All-space antenna array for detection of UAV in urban area.

If A and B are RF voltages measured at the reference boresight and incident antennas, respectively, then

$$A = M \sin (\omega t) \tag{8.2}$$

and

$$B = M \sin (\omega t + \psi) = M \sin (\omega t - \sin \varphi) \tag{8.3}$$

where M is a common constant defined by signal power. This shows that the angle of arrival φ is contained in the RF argument or phase difference of the two beams for all signals off the boresight axis. Direction finding by way of amplitude comparison methods can provide a root mean square (RMS) accuracy smaller than 2° in 100 ns after a direct wave arrives [3]. High accuracy phase measurements provide fast and accurate direction finding. The monopulse method is capable of providing critical information about targets' position, speed, and identity within a small time frame, typically less than 1 millisecond.

An array of staring directional antennas, as shown in Figure 8.7, can fulfill the requirements for continuous detection and tracking of multiple targets

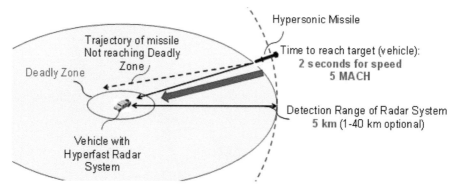

Figure 8.8   Diagram demonstrating the application of directional antenna array for the detection of hypersonic missiles. Doppler shifted frequency components allow to exclude missiles not pointed to deadly zone. Detection range is determined by transmitted power and does not have a minimum limit.

simultaneously. The proposed antenna array is designed for army vehicles and can be applied in various military applications [4–10].

The all-space antenna array is capable of simultaneously detecting multiple targets within a range of 5 km (with an optional range of 1–40 km). The detection time is faster than 0.2 seconds, allowing for quick identification and response. The identification of targets can be achieved by analyzing their frequency spectrum signatures. This antenna array is specifically designed for protecting airborne, ground, and sea surface vehicles against hypersonic missiles and drones. Hypersonic missiles, which can travel at speeds of Mach 5, can reach a target from a distance of 5 km in approximately 2 seconds (as illustrated in Figure 8.8). By analyzing the Doppler shifted frequency components, the directional antenna array can exclude missiles that are not aimed at the protected zone. The detection range is determined by the transmitted power and is not limited by a minimum distance. Figures 8.9–8.12 depict the application of the proposed directional antenna array for the protection of airborne, ground, and naval vehicles against hypersonic missiles and drones. The array enables effective detection and identification of these threats, enhancing the overall defense capabilities of the vehicles.

8.4 Conclusion

In conclusion, the proposed tactical radar system offers several advantages for the fast and automatic detection and identification of multiple hypersonic and low cross-sectional targets. The system utilizes transceiver modules

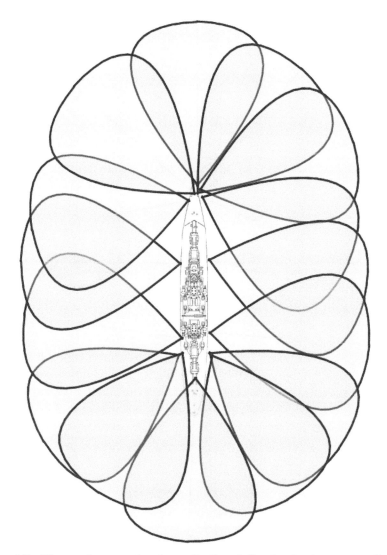

Figure 8.9 Diagram demonstrating the application of directional antenna arrays for warships. Antenna arrays cover the entire sky and do not need time for scanning the sky or switching sectors of observation. The distribution of directional antennas provides a green zone free from electromagnetic irradiation inside a warship and additional protection for the radar (communications) system. An antenna array can be multi-frequency and multi-functional.

distributed around the perimeter of the protected vehicle, ensuring continuous and simultaneous coverage of the entire 360° search space.

By employing multiple directional antennas in the transceiver modules, the system achieves continuous detection and tracking of multiple targets

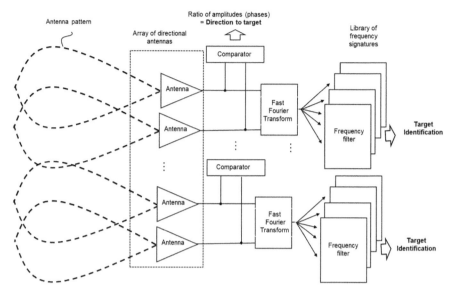

Figure 8.10 Block diagram of a monopulse directional antenna array module for a staring antenna array.

with higher signal gain compared to phase arrays. The distribution of transceiver modules reduces radar vulnerability, as each module covers a specific sector and is protected from external electromagnetic pulses by the directional antennas.

The system utilizes the monopulse method, which allows for very fast signal processing using ratios of amplitudes, phases, or frequency components. This enables rapid target identification in a matter of microseconds. The non-scanning monopulse system reduces the need for high transmitting power while increasing the radar range by integrating a larger number of signals compared to scanning radar systems. Additionally, the monopulse method provides significantly improved target resolution compared to scanning radars.

Overall, the proposed radar system offers a highly efficient and effective solution for detecting and identifying hypersonic and low cross-sectional targets, providing enhanced situational awareness and protection for airborne, ground, and naval vehicles.

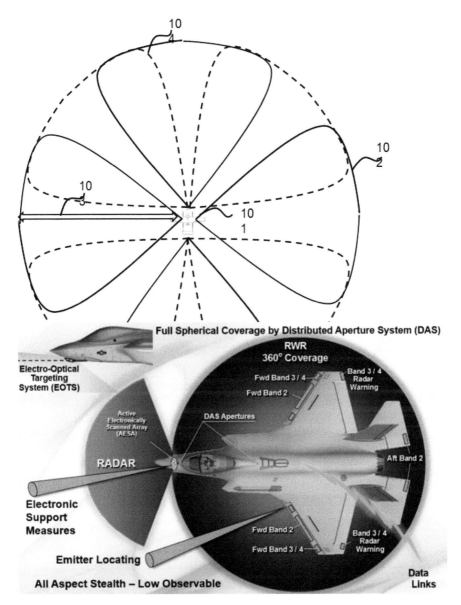

Figure 8.11 Diagram demonstrating the application of a directional antenna array for air-crafts. An antenna array provides omnidirectional simultaneous covering.

Figure 8.12 Distributed antenna modules can be made as Slat armor for additional vehicle protection against missiles or anti-tank rocket-propelled grenade attacks.

References

[1] Margot van Loon, Larry Wortzen, Mark B. Schneider, "Hypersonic Weapons", American Foreign Policy Council Defense Technology Program Brief, No. 18, May 2019.

[2] Theresa Hitchens, "Pentagon needs to prioritize hypersonic defense, not offense. Defense industry news, analysis and commentary. February 07, 2022.

[3] S. Lipsky, "Microwave passive direction finder", SciTech Publishing Inc. Raleigh, NC 27613, 2004.

[4] P. Molchanov, A. Gorwara, "Fly Eye radar concept". IRS2017. International Radar Symposium, Prague, July 2017.

[5] P. Molchanov "Tactical radar system for detection of hypersonic missiles and UAS", patent appl. 17/651,800 02/02/2022.

[6] P. Molchanov "Passive radar system for detection of low profile low altitude targets", US patent appl. 17/971,582, 10/22/2022.

[7] P. Molchanov "Multi-beam multi-band antenna array module" US patent appl. 17/971,616, 10/23/2022.

[8] P. Molchanov "Multi-beam multi-band protected communication system" patent appl. 17/740,581, 05/10/2022

[9] P. Molchanov, A. Gorwara, "Fly Eye radar. Detection through high scattered media", (PDF) Fly Eye radar: detection through high scattered media (researchgate.net)

[10] A. Gorwara, P. Molchanov, "Fly Eye Radar Concept" (PDF) All-digital radar architecture (researchgate.net)

9

All-space Multi-beam Staring Communication Systems

9.1 Challenges of All-space Communications

The challenges of all-space communication systems are as follows:

1. Limited coverage: Mechanical scanning and phase array electronically scanning antenna arrays have limitations in covering the entire sky simultaneously. Phase arrays, in particular, have phase control errors on the array edges, and the number of antenna elements increases these errors. Even omnidirectional antenna elements do not provide omnidirectional coverage, leading to coverage gaps (Figure 9.1) and size of beamforming circuits.

2. Decreased gain and link budget: To achieve simultaneous coverage of the entire sky, multiple beams can be formed using phase arrays. However, each additional steering beam within the phase sub-array decreases the maximum possible gain and range of communication. The number of switching communication links is also proportional to decreasing the link budget, which affects the non-interrupting time of connection with each satellite (Figure 9.2).

3. Scanning time and beamwidth tradeoff: Scanning time for each beam in a phase antenna array is inversely proportional to the beamwidth. Smaller beamwidths result in longer scanning times to cover the entire observation area and find satellites for connection. On the other hand, larger beamwidths decrease scanning time but sacrifice directional accuracy and gain/maximum connection range (Figure 9.3).

These challenges highlight the need for innovative solutions to overcome limitations in coverage, gain, link budget, scanning time, and directional accuracy in all-space communication systems. Addressing these challenges will be crucial for achieving efficient and reliable communication in the vast and dynamic space environment.

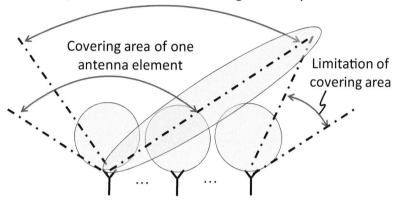

Figure 9.1   Limitation of covering area in a planar phase array. Even omnidirectional antenna elements do not provide omnidirectional covering because of phase errors on the edge of antenna patterns. More antenna elements lead to increasing phase errors.

The challenge of future mobile antenna array technology for all-space is to make "magic bullet" for all orbits communication. Formalized requirements to what antenna arrays must do and how it can do it for all-space MEO, LEO, and VLEO (middle Earth orbit, low Earth orbit, and very low Earth orbit) communications and possible navigation are:

1.   Simultaneous coverage: The antenna array should be able to simultaneously cover the entire sky to detect and receive signals from all possible sources.

2.   Continuous staring: For maximum efficiency, sensitivity, and range, the antenna array must provide continuous staring, without the need for scanning or switching, for each satellite or signal source. This ensures non-interrupting communication and maintains the link budget.

3.   Simultaneous signal processing: The antenna array should be capable of simultaneous processing of signals from all satellites or signal sources. It should have multiple channels and optionally support multi-frequency and multi-function operation.

4.   Solid-state and protection: The antenna array should be solid-state with no moving parts, providing protection against jamming, spoofing, and electromagnetic pulses (EMPs).

PHASE ARRAY

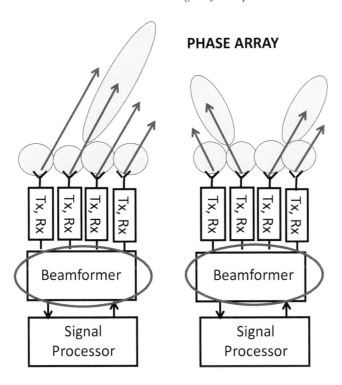

Figure 9.2 Dividing of one steering beam to two will decrease gain in each beam twice if the antenna aperture is same for one or two beams.

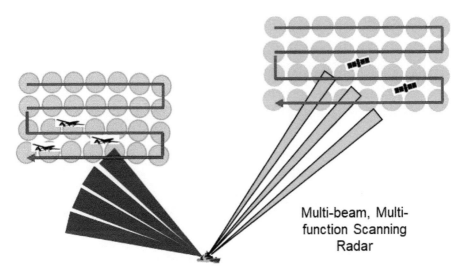

Multi-beam, Multi-function Scanning Radar

Figure 9.3 Multi-beam electronically scanning antenna array. Smaller beamwidth leads to increasing scanning time for the same scanning area.

While multi-beam antenna arrays can cover the entire sky, traditional phase arrays with steering beams have limitations in providing non-interrupting communication with all signal sources simultaneously. Scanning and switching beams also affect the communication line's budget. To overcome these challenges, a nature-inspired antenna array architecture is proposed, comprising multiple staring directional antennas.

Nature inspired antenna array architecture with multiple staring directional antennas will cover the entire sky and allows virtual scanning and real-time recording of digital holograms. Integration of a holographic non-scanning antenna system [1, 2] with a monopulse high-directional accuracy method of fast signals processing directly on multi-beam multi-axis overlap antennas [3] and direct digitizing of received signals will provide all-sky communication, fast threat detection, tracking, and recognition.

Each of angular tilted directional antennas (sensors) must be coupled with a digitizer (ADC and SDR) and connected to a processor by digital interface [4–10]. An antenna array comprises multiple directional antennas that can be closely positioned or loosely distributed (like spider eyes) along the perimeter of a ground vehicle or aircraft, connected with receiving/transmitting front-end circuits and connected to the processor by a digital interface.

Direct digitizing of received signals relative to common synchronization (sampling) source directly on each directional antenna allows to create a real-time digital hologram and distribute antennas by a digital interface.

Transformation and processing of received signals in time domain, frequency domain, and multi-axis space domain provides additional possibility to enhance reliability and communication quality or target recognition for radars. This architecture integrates holographic radar staring antenna systems, monopulse high-directional accuracy methods, and direct digitization of received signals. Each angular tilted directional antenna is coupled with a digitizer and connected to a processor through a digital interface. The antenna array can be closely positioned or loosely distributed around the perimeter of a ground vehicle or aircraft, and the received signals are digitized in real time.

A. Simultaneously Covering the Entire Sky

The proposed antenna array architecture includes the concept of overlapping antenna patterns in one or multiple directions (multi-axis). This allows for enhanced coverage and reception of signals from multiple sources simultaneously. Here are some key points regarding the antenna patterns:

1. One-directional overlap: In this configuration (Figure 9.4), the antenna patterns overlap in one direction, typically along the X-axis. This

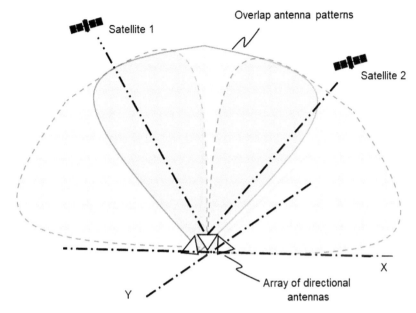

Figure 9.4 Ground-based all-space antenna array with *X*-axis overlap antenna patterns.

configuration can be employed in ground-based all-space antenna arrays. The overlapping patterns ensure that signals from different sources within the coverage area can be received simultaneously, maximizing the information obtained.

2. Two-directional overlap: The antenna patterns can also overlap in two directions, such as along the *X*- and *Y*-axes (Figure 9.5). This configuration is suitable for airborne carrier-based all-space antenna arrays. The overlapping patterns in two directions provide even greater coverage and reception capabilities, enabling the array to receive signals from multiple sources with high efficiency.

3. Multi-axis overlap: The antenna patterns can be designed to overlap in multiple directions, forming a multi-axis coverage (not shown in the figures). This configuration further enhances the information obtained from signal sources and ensures reliable reception with a high data rate. Multi-axis overlap allows for comprehensive coverage in three dimensions, enabling the array to capture signals from various angles and positions.

These overlapping antenna patterns, combined with the staring, non-scanning continuous wide coverage capability, provide significant advantages in

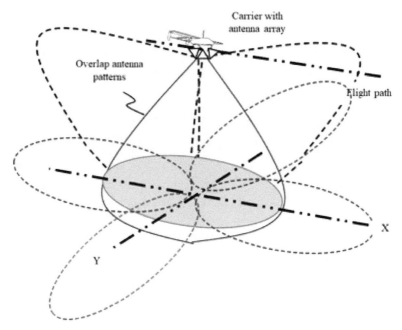

Figure 9.5   Airborne carrier-based all-space antenna array with two-directional *X*- and *Y*-axis overlap antenna patterns.

all-space communication systems. They enable the antenna array to receive maximum information from all signal sources simultaneously and optimize the line budget for efficient and reliable communication. Whether deployed on ground-based, airborne, or sea-based vehicles, this antenna array architecture offers enhanced reception capabilities for all-space communication and can support high data rates.

B. Non-interrupting Line Budget

Planar phase array antennas with isotropic elements have limitations in terms of the covered space sector due to phase errors on the edges of the antenna patterns. Additionally, scanning beams in a phase array antenna system require a significant amount of time to scan the entire coverage sector. This limits the communication time and line budget available for each satellite, resulting in very short connection times and a reduced link budget.

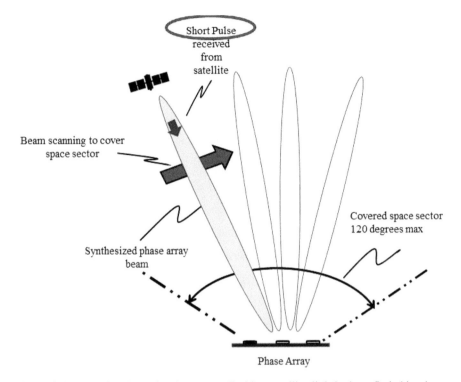

Figure 9.6 Scanning beam in phase array limiting satellite link budget. Switching beam between multiple satellites making link budget dramatically worse.

While a multi-beam antenna system can detect and track multiple satellites, the number of beams decreases the gain of each individual beam within the antenna array. The gain in one direction remains constant for a phase array with a fixed aperture, but as the beams are divided and shared among multiple directions, the gain decreases. This reduction in gain negatively affects the sensitivity and range of the antenna array.

Moreover, the scanning beams in a phase array antenna system limit the active or illuminating time available for communication with satellites. This further reduces the connection time and diminishes the link budget, as depicted in Figure 9.6.

These challenges highlight the limitations of traditional phase array antennas with scanning beams in terms of coverage, connection time, and link budget. Alternative antenna array architectures, such as the proposed multi-beam multi-axis overlap antenna patterns with staring capabilities, aim

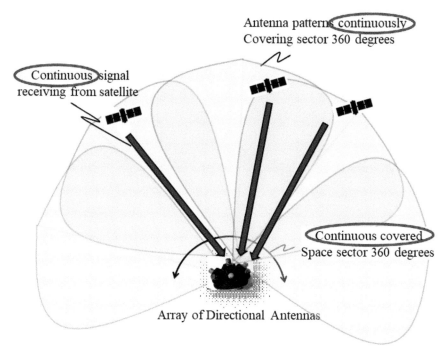

Figure 9.7   Array of directional antennas can cover all required space, providing high gain in any direction for multiple beams and non-interrupting illuminating all targets or communication with multiple satellites with non-interrupting full link budgets.

to address these limitations and provide enhanced performance in all-space communication systems.

A communication diagram of the proposed staring array of directional antennas for simultaneous communication with LEO/MEO/GEO satellites is presented in Figure 9.7. Plurality of directional antennas provides continuous communication with multiple satellites and maximum satellite link budget in wide space area (all sky).

In contrast, in phase array, the antenna elements need to be switched for observation of different space sectors and/or multi-band functions. It leads to decreasing time for communication with one satellite (link budget). Electronically steerable (means phase array) panel antenna also limit the frequency band of communication link. An array of directional antennas allows to solve these problems.

C. All-Space Antenna Array Applications

The best way to protect your message from an enemy is if the enemy will not receive it. Do not send your messages by isotropic communication to

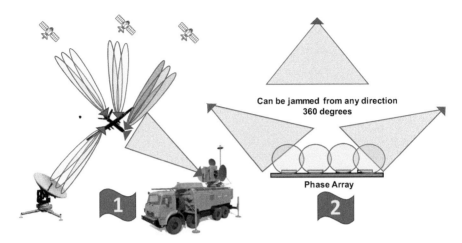

Figure 9.8 All-space communication antenna array distributed around the UAV perimeter provides full performance links and additional protection against jamming, spoofing, and EMP signals.

anywhere! Do not receive messages and jamming and spoofing signals by isotropic communications from anywhere.

Even an adaptive phase array with omnidirectional antenna elements cannot be reliable in protecting your communication because all antenna elements in a planar phase array are omnidirectional and receive all signals useful and jamming simultaneously.

The proposed fly eye structure approach using directional antenna arrays and limiting the transmission and reception of signals to specific sections of space can indeed enhance the protection of communication. By avoiding isotropic communication that broadcasts messages everywhere, the risk of interception, jamming, and spoofing by unauthorized parties can be significantly reduced.

In the proposed communication antenna array (Figure 9.8), the entire sky is covered, allowing connections with all signal sources (customers) simultaneously. The use of directional antennas ensures that transmissions are focused only on the intended recipients, providing a level of protection for communication information. Multiple staring antenna arrays enable continuous and non-interrupting connections with each signal source, allowing for full utilization of the line's budget.

Simultaneous monopulse processing of signals from all signal sources enables fast and efficient multi-beam communication. Direct digitization of signals on the antennas provides a wider bandwidth and enables the distribution of antennas around the UAV, offering flexibility in deployment.

The antenna array can also be designed to be multi-frequency and multi-mode, further enhancing its capabilities. By providing 360° full/hemisphere coverage without the need for scanning or switching, the antenna array ensures continuous and uninterrupted communication. The use of automatic gain control for each beam allows for adaptability in different environments, such as urban or mountainous areas.

Distributing the antenna array around the perimeter of a vehicle or connecting it to a network can provide additional protection against jamming, spoofing, and electromagnetic pulse (EMP) signals, as depicted in Figure 9.8.

Overall, the proposed directional antenna array architecture offers improved communication protection by minimizing the exposure of messages to unwanted recipients and mitigating the risks associated with interception, jamming, and spoofing.

An important notice about the safety of communication systems is that directional communications have a few important safety parts. Here are the key aspects:

1. Directional communication: Directional communication involves the use of narrow, accurately directed beams for specific connections, as shown in Figure 9.8. By using directional antennas, the communication signals are confined to specific paths, reducing the risk of interception or unintended reception.

2. Distribution of directional antennas: To further enhance the safety of communication systems, it is advisable to distribute directional antennas around the carrier or vehicle. Figure 9.9–9.11 illustrates the distribution of antennas around the perimeter of a vehicle, leveraging its body for additional protection against electromagnetic interference or pulses (EMP). These antenna array modules can be integrated into the vehicle's structure, serving a dual purpose of communication and protection.

3. Protection against interference: Directional antennas should also be safeguarded against interference, jamming, and spoofing signals originating from various directions.

Conventional planar phase antenna arrays, utilizing omnidirectional antenna elements and beamformers, can be vulnerable to signals coming from any direction. When subjected to simultaneous jamming signals, the phase control circuits may become overwhelmed before the system has a chance to process them or form nulls to counteract the jamming direction.

Taking these factors into account, the design and implementation of a communication system should prioritize directional communication, the

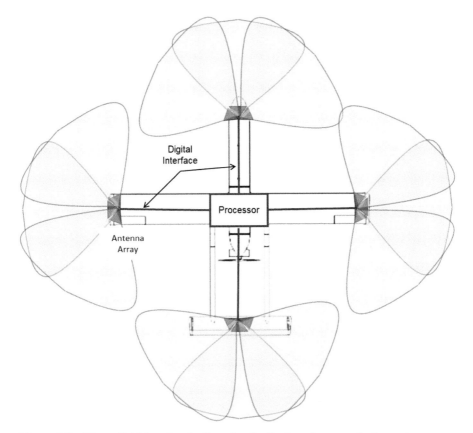

Figure 9.9 Direct digitizing signals allow the distribution of communication antenna array around an UAV perimeter. Distributed staring antenna arrays provide all-sky communication and enhanced protection.

strategic placement of directional antenna arrays, and measures to protect the antennas from interference. By adopting such precautions, the safety and security of the communication system can be significantly enhanced.

References

[1] S. A. Harman, "A Comparison of staring radars with scanning radars for UAV detection" Proc. of the 12th European Radar Conference, Paris, France, Sept 2015.

[2] S. A. Harman, "Holographic radar development", Microwave Journal, Vol. 2, February, 2021, www.microwavejournal.com/articles/35410-holographic-radar-development.

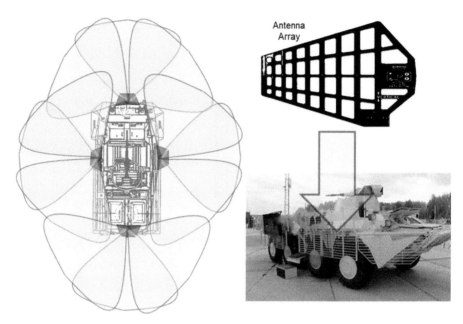

Figure 9.10   A distributed communication/radar/navigation antenna array can provide additional fire protection for vehicles.

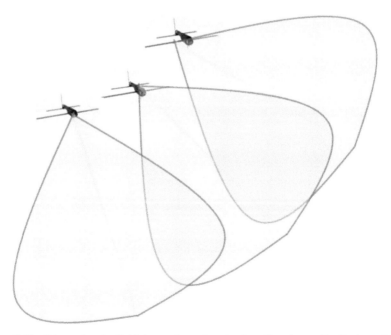

Figure 9.11   Direct signals digitizing and synchronization allow loose distribution of communication/navigation system between swarms wirelessly.

[3] S. E. Lipsky, "Microwave passive direction finder", SciTech Publishing Inc. Raleigh, NC 27613, 2004.

[4] P. A. Molchanov, A. Gorwara, "Fly Eye radar concept". IRS2017. International Radar Symposium, Prague, July 2017.

[5] A. Gorwara, P. Molchanov. Multibeam Monopulse Radar for Airborne Sense and Avoid System, Proc. of 2016 SPIE Remote Sensing and Security+Defense Conference, Edinburgh, UK, Paper #9986-3, September 2016.

[6] P. A. Molchanov, O. V. Asmolova, "All-digital radar architecture", Conference: SPIE Security + Defense, DOI: 10.1117/12.2060249, October 2014.

[7] P. A. Molchanov, M. V. Contarino, "Multi-beam antenna array for protecting GPS receivers from jamming and spoofing signals" US Patent US20140035783, 2014.

[8] P. A. Molchanov "Tactical radar system for detection of hypersonic missiles and UAS", US Patent appl. 17/651,800 02/02/2022.

[9] P. A. Molchanov "Multi-beam multi-band protected communication system" US Patent appl. 17/740,581, 05/10/2022.

[10] P. A. Molchanov "Passive radar system for detection of low-profile low altitude targets", US Patent appl. 17/971,582, 10/22/2022.

10

Fast Synthetic Aperture Radar

10.1 Fast Synthetic Aperture Radar

The synthetic aperture radar (SAR) is a radar technique used to create high-resolution images of objects, such as landscapes. SAR systems utilize the motion of the radar antenna, whether physical or synthetic, to generate a large effective antenna aperture and achieve a finer spatial resolution. In SAR, the radar antenna moves over a target region while emitting and receiving radar signals. The distance traveled by the SAR device during the illumination of the target scene determines the synthetic aperture size, which is a measure of the effective antenna size. A larger synthetic aperture leads to higher image resolution, allowing SAR to capture detailed images using relatively small physical antennas.

One of the advantages of SAR is that it maintains a consistent spatial resolution over a range of viewing distances. Objects that are further away from the radar antenna remain illuminated for a longer period of time, effectively creating larger synthetic apertures for these distant objects. This property helps SAR systems produce images with consistent resolution regardless of the target's distance from the radar. Also an inverse technique called inverse synthetic aperture radar (ISAR) is proposed where the target itself is in motion, creating a similar effect of a moving antenna. In ISAR, the target's motion is used to generate the synthetic aperture and produce high-resolution images.

Overall, SAR is a valuable radar technique for obtaining detailed images of objects and landscapes, offering advantages in terms of resolution and viewing distance. The fly eye radar based on receiving larger amount of information about targets and fly eye concept can be applied for enhancing image resolution in SAR imaging radar. The fly eye imaging radar has a few differences compared to SAR and ISAR radars (Figure 10.1). SAR radars receive signals transmitted from one direction and recreates images by gathering information received from different positions of the platform (UAS). Each picture consists of only two-dimensional information about targets

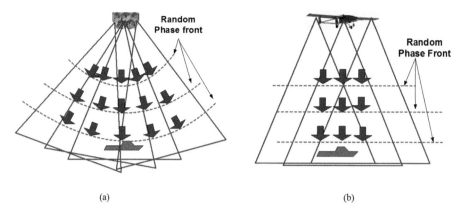

(a)   (b)

Figure 10.1 Fly eye antenna array allows to record real-time digital hologram. A large number of transmissions in the direction of the target and received hit target pulses can enhance the target recognition and enhance the resolution of imaging radar. Staring directional antennas can be positioned at one point (a) or distributed over the perimeter of UAS (b). Note that a digital hologram is recorded by receiving signals from all areas covered by a set of overlapped beams in a few directions and processing them simultaneously by multi-channel monopulse processors. This virtual holographic digital scanning allows to receive more information from the target and processes it much faster than a regular SAR.

received from one point of view. A combination of signals, received from different points of view, allows to enhance the target image.

Radar signals transmitting at the same time from all directional antennas create a random phase front, and signals received by each antenna consist of signals reflected from the target from different directions simultaneously. Each received signal consists of information about all targets like hologram and can be applied for reconstruction of a target image. The system does not need a phase processor, and a random phase front can be applied [2].

A monopulse synthetic aperture radar (SAR) system that utilizes multiple directional antennas with overlapping antenna patterns is positioned in quadrature or multi-axis directions. This configuration allows for wide coverage of the space sector and enables simultaneous fast signal processing from all directions. In this system, the transmitting antennas generate radar signals with a random phase front, while the receiving antennas capture the signals reflected from the target from various directions. Each received signal contains information about the target, similar to a hologram, which can be used to reconstruct the target image.

The monopulse technique, combined with multi-channel digital processing, is applied to the received signals in separate receiver chains coupled with a digital multi-channel processor. By processing the amplitudes, phases, and frequency components of the signals relative to those in the set of over-lapping beams, the system can achieve higher imaging resolution and suppress the influence of medium and clutter. The array of directional antennas can be designed to operate in multi-frequency and multi-mode regimes.

Compared to regular SAR systems with scanning antennas, the proposed staring antenna arrays offer advantages in terms of observation speed and image resolution. Scanning antennas require time to observe a wide area, which can limit the speed of the SAR carrier vehicle. By utilizing staring antenna arrays and virtual holographic recording of wide spot information simultaneously, the system can increase its speed. The overall time of signal processing will depend on the speed of the processor employed in the system.

Regular SAR with scanning antenna needs some time for the observation of a wide area and may limit the speed of an SAR carrier vehicle. The size of the beam spot (proportional to wavelength) must be smaller for better image resolution. As result, the scanning time will be increased and the frequency will be higher, with loosing signal penetration. Stephen A. Harman explained that staring antenna arrays are better than scanning antennas for these or similar applications and proposed a holographic system for simultaneous observation of wide area [3, 4]. Virtual holographic recording of wide spot information will simultaneously increase the speed of the system. The time of signal processing will depend on the processor speed.

Stephen Lipsky explained how to make better directional accuracy for an antenna system by using the monopulse method consisting of simultaneous multi-channel processing signals in a set of overlap beams. The application of lower-frequency overlapping antennas provides a wide cover area and, at the same time, a better directional accuracy, image resolution, and signal penetration [5]. Direction finding by way of amplitude comparison methods can provide a root mean square (RMS) accuracy smaller than 2° in 100 ns after a direct wave arrives. High-accuracy phase measurements provide high accuracy and fast direction finding. But most importantly, the monopulse method does not require long time, from milliseconds for small amount operations to tens of seconds for FFT (Fast Fourier Transform) computer calculations, and can provide critical information about targets position, speed, and identity.

The fly eye radar system combines the virtual recording of a digital hologram with monopulse simultaneous signal processing using a set of directional antennas with overlapping antenna patterns. By directly digitizing the signals on the antennas and synchronizing them with a central computer or sampling system, the system can create a fast synthetic aperture radar (SAR) that enables simultaneous object observation in multiple directions.

In this configuration, the antennas' patterns overlap in two directions, typically denoted as *X* and *Y*. This overlapping arrangement allows for improved imaging resolution compared to traditional scanning SAR systems. The virtual recording of the digital hologram captures the information from multiple directions, enabling a comprehensive view of the target or scene (Figures 10.2, 10.3).

By employing monopulse simultaneous signal processing techniques, the fly eye radar system can extract and analyze the signals received from the directional antennas. This processing methodology, combined with the overlapping antenna patterns, enhances the imaging resolution and provides a fast and accurate SAR capability. It is important to note that the fly eye radar system's performance and effectiveness would depend on various factors, including the specific design, antenna characteristics, signal processing algorithms, and environmental conditions [6].

The antenna array configuration depicted in Figure 10.4 shows the potential distribution of antenna transceiver modules between unmanned aircraft system (UAS) swarms. By distributing the antenna transceiver modules across multiple UAS, the system can extend the simultaneous area of observation.

This distribution allows for wider coverage and the ability to observe and collect data from a larger geographical area simultaneously. Each UAS within the swarm would be equipped with its own set of antenna transceiver modules, which can communicate and coordinate with each other to enhance the overall performance and capabilities of the system. The overlapping antenna patterns, as illustrated in Figure 10.4, provide additional information for object imaging and recognition. The simultaneous coverage and holographic digital/virtual parallel monopulse processing of signals from all antennas enable fast and efficient information processing in the SAR system (Figures 10.5–10.7).

Furthermore, the reference antennas within the array can be utilized for interference signal and clutter suppression. By leveraging signals from these reference antennas, the system can distinguish between desired signals and unwanted interference or clutter, thereby improving the accuracy and reliability of the imaging and recognition process. Specific implementation and performance of such a system would depend on various factors, including the number and arrangement of antenna transceiver modules, the signal processing algorithms employed, and the capabilities of the UAS swarm.

The application of the SAR radar technology for maritime vessel detection on the open sea involves optimizing the performance characteristics of the radar system. To achieve this, proper design considerations are made for

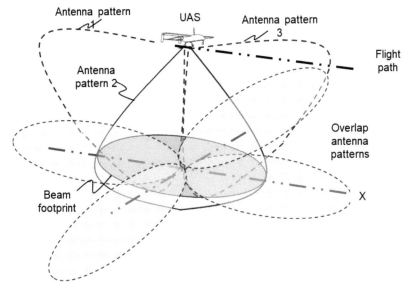

Figure 10.2 Fly eye fast SAR with enhanced imaging resolution. Antenna patterns are over-lapping in two X and Y directions.

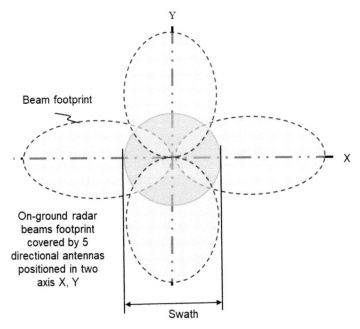

Figure 10.3 On-ground beams footprint for Fly eye fast SAR with enhanced imaging resolution.

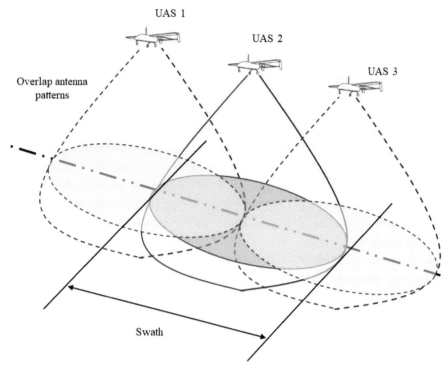

Figure 10.4   Antenna transceiver modules can be distributed between UAS swarms to extend a simultaneous area of observation.

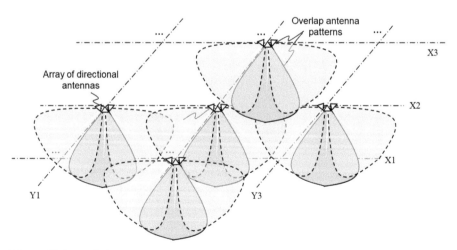

Figure 10.5   A digital hologram can be recorded faster and in a wider area by distributing arrays of directional antennas in two or more axes.

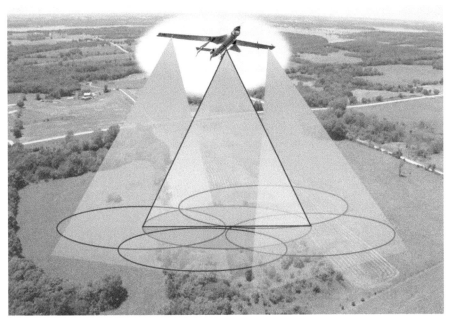

Figure 10.6 Distribution of antennas with overlapping antenna patterns providing fast wide area virtual scanning, additional information from a few directions, and reference signals for clutter suppression and additional jam and spoof EMP protection.

airborne radar operations, taking into account various interdependent system parameters.

One aspect of enhancing radar system performance is the exploration of limitations and tradeoffs among different system parameters. Through careful analysis and optimization, parameters can be identified that offer the best overall performance for maritime SAR applications. Additionally, various technologies have been proposed to address specific challenges in radar imaging, such as the apodization of spurs in radar receivers using multi-channel processing. Spurious energies or signals in radar data can arise from non-ideal behavior of components and circuits within the system. Examples of such non-ideal behavior include I/Q imbalance, nonlinear component characteristics, and additive interference (Figures 10.5–10.7).

By mitigating the influence of undesired spurs or spurious energy, the radar system can improve the identification and recognition of true targets in the radar image. This suppression of unwanted spurs allows for enhanced clarity and accuracy in target detection. It is important to recognize that a radar system is composed of multiple components, each of which may

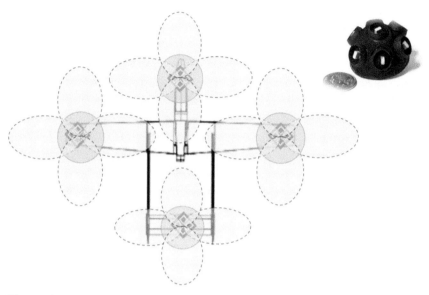

Figure 10.7   Sample of distribution of directional antenna modules around the perimeter of UAS.

exhibit some degree of non-ideal behavior. These non-idealities can impact the purity and quality of the signal being processed by the radar system. Therefore, efforts are made to identify and mitigate these non-ideal effects to optimize the system's performance and ensure accurate target identification in radar imaging. The design and optimization of radar systems for maritime SAR involves a comprehensive analysis of various parameters and the implementation of techniques to minimize non-idealities and enhance the detection and identification of targets in radar images [7–10].

Figure 10.8 represents a schematic of a transceiver antenna module designed for fast enhanced image resolution in SAR (synthetic aperture radar) systems. The module consists of a set of non-scanning transmitting and receiving antennas with overlapping antenna patterns. These antennas are positioned in quadrature or multi-axis directions, covering a wide space sector.

Each receiving antenna is coupled to a separate receiver chain and a multi-channel processor. This configuration allows for the simultaneous application of monopulse and digital multi-axis multi-channel processing of all signals received by the antennas. By utilizing the monopulse method and introducing shifts in amplitudes, phases, and frequency components of the received signals in relation to signals in the overlap receiving antenna beams,

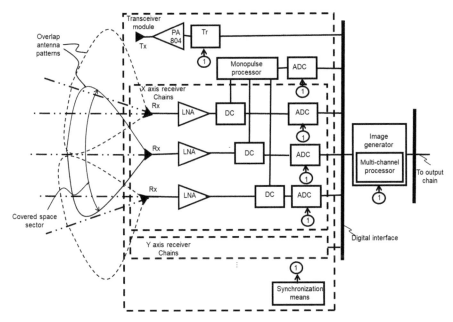

Figure 10.8   Schematic of transceiver antenna module for faster enhanced image resolution SAR.

a higher imaging resolution can be achieved. This approach also helps in suppressing the influence of media and clutter, thereby improving the quality of the radar image. The transceiver antenna module shown in Figure 10.8 represents a system design aimed at optimizing the imaging capabilities of SAR systems. By leveraging the benefits of multi-channel processing and monopulse techniques, the module enables faster signal processing and enhanced image resolution, which is crucial for accurate target identification and clutter reduction in SAR applications [11–15].

Here are the key points:

- The monopulse synthetic aperture radar (SAR) system is designed for fast, high-resolution imaging of ground and/or airborne objects.

- The radar system comprises both transmitting and receiving non-scanning means organized as monopulse transceiver modules.

- Each monopulse transceiver module is responsible for covering at least one specific space sector.

- Each module consists of a monopulse processor, at least one transmitter connected to at least one transmitting antenna for the respective space sector, and a set of non-scanning receiving antennas with overlapping antenna patterns in quadrature or multi-axis directions, also covering the same space sector.

- The receiving antennas are coupled to separate receiver chains, which are further connected to the monopulse processor.

- The monopulse processor performs simultaneous (monopulse) multi-axis processing of all signals received by the antennas, taking into account the amplitudes, phases, and frequency components shift of the signals in the receiving antennas with overlapping patterns.

- The outputs of the monopulse processor and receiver chains are connected to an image generator through a digital interface.

- The image generator incorporates a quadrature or multi-axis multi-channel processor that simultaneously processes all signals from the receiver chains, considering the amplitude, phase, and frequency components shift of the signals in the receiving antennas, and generates the radar image.

- All transmitters, receiver chains, and processing means are connected with synchronization means to ensure proper coordination.

- The transceiver modules, image generator, and radar output chain can be connected using microwave and/or fiber optic digital interfaces.

- The transceiver modules can be positioned on unmanned aerial systems (UAS) or other airborne or ground carriers. Multiple modules can be distributed around a single carrier, or multiple modules can be distributed among a swarm of carriers connected to a radar communication network.

- The transmitting, receiving, and processing means can be arranged to handle simultaneous transmission, reception, and processing of signals on different frequencies (multi-frequency signals), using separate antennas and filters in each receiving and processing chain. Alternatively, they can be arranged to handle simultaneous transmission, reception, and processing of different modes (multi-mode signals), such as communication, navigation, and control, using separate antennas, filters, and processing means in each transmitter, receiver, and processing chain.

This configuration enables the SAR system to perform fast and high-resolution imaging by utilizing monopulse processing, multi-axis coverage, and simultaneous processing of signals from multiple antennas. The system can be adapted to different carrier platforms and can handle various frequencies and modes of operation.

10.2 Conclusion

The utilization of simultaneous multi-channel holographic direct digitizing signals from a wide covered area enables the fast and high-resolution imaging capabilities of the radar system.

The combination of a staring array of directional antennas and fast signal processing allows for significantly longer object illuminating and information gathering, with a potential improvement of over 1000 times compared to other systems.

The simultaneous monopulse processing of signals with multi-axis overlap antenna patterns can enhance the image resolution by a factor of 10–1000, providing a clearer and more detailed representation of the targets. Additionally, the inclusion of signals from reference directional antennas helps in suppressing clutter, noise, and mitigating the influence of the medium.

Directly digitizing signals on the antennas allows for the distribution of antennas and offers protection against jamming and spoofing signals. This flexibility enhances the overall robustness and reliability of the system.

The simultaneous multi-channel signal processing from each antenna enables the detection of fast-moving targets, accommodating higher carrier speeds without sacrificing accuracy and performance.

Furthermore, the implementation of multi-axes overlap antenna patterns contributes to a wider coverage area and provides additional information about the targets, leading to improved target recognition and image resolution.

Lastly, the use of directional antennas with direct digitizing capabilities allows for the distributed placement of antennas around the carrier perimeter, offering more flexibility and adaptability to different system configurations.

References

[1] Wikipedia, Synthetic Aperture Radar, SAR
[2] P. A. Molchanov, O. V. Asmolova, "All-digital radar architecture", Conference: SPIE Security + Defense, DOI: 10.1117/12.2060249, October 2014.

[3] S. A. Harman,"A Comparison of staring radars with scanning radars for UAV detection" Proc. of the 12th European Radar Conference, Paris, France, Sept 2015.

[4] S. A. Harman,"Holographic radar development", Microwave Journal, Vol. 2, February, 2021, www.microwavejournal.com/articles/35410-holographic-radar-development.

[5] S. E. Lipsky, "Microwave passive direction finder", SciTech Publishing Inc. Raleigh, NC 27613, 2004.

[6] P. A. Molchanov, A. Gorwara, "Fly Eye radar concept". IRS2017. International Radar Symposium, Prague, July 2017.

[7] Armin Walter Doerry, Sandia National Laboratories, "Performance Limits for Maritime Wide-Area Search (MWAS) Radar", Technical Report · May 2020.

[8] Armin W Doerry, Douglas L Bickel, "Antenna Requirements for GMTI Radar Systems", Sandia Report Sand2020-2378, Printed February 2020

[9] D. L. Bickel, A. W. Doerry, "On minimum detectable velocity," Proc. SPIE 11003, Radar Sensor Technology XXIII, 110030N (3 May 2019); doi: 1117/12.2518121, SPIE Defense + Commercial Sensing, Baltimore, Maryland, United States, 2019.

[10] Armin W. Doerry, Douglas L Bickel, Radar Velocity Determination Direction of Arrival Measurements, US patent 9,846,229 B2, Dec.19, 2017.

[11] P. Molchanov "Monopulse synthetic aperture radar", US patent appl. 17/699,129, 03/20/2022.

[12] A. Gorwara, P. Molchanov. Multibeam Monopulse Radar for Airborne Sense and Avoid System, Proc. of 2016 SPIE Remote Sensing and Security+Defense Conference, Edinburgh, UK, Paper #9986-3, September 2016.

[13] P. A. Molchanov, M. V. Contarino, "Multi-beam antenna array for protecting GPS receivers from jamming and spoofing signals" US Patent US20140035783, 2014.

[14] P. A. Molchanov "Tactical radar system for detection of hypersonic missiles and UAS", US Patent appl. 17/651,800 02/02/2022.

[15] P. A. Molchanov "Passive radar system for detection of low-profile low altitude targets", US Patent appl. 17/971,582, 10/22/2022.

11

Detection of Hazardous Materials and Concealed Objects

11.1 Introduction

The fly eye antenna array offers significant advantages in the detection, identification, and visualization of various materials and concealed objects. One key advantage is that it can cover a wide area of the sky simultaneously using relatively low radio frequency signals. At first, this allows for the use of a smaller number of antennas compared to phased antenna arrays, typically ranging from 10 to 100 times fewer antenna elements. Second, low radio frequencies have good penetration through walls, ground, and human body and do not requiring high-power danger for human transmitting energy.

The monopulse processing of signals in directional antennas with overlap antenna patterns enables high-accuracy direction finding, which is crucial for applications requiring precise target localization. The multi-channel direct signal processing in both the space–time and frequency domains facilitates fast detection and recognition of materials and/or objects.

Given these advancements, the fly eye antenna array can be recommended for a wide range of applications. While specific applications are presented in the author's paper, it can be utilized in various fields that benefit from its capabilities, such as surveillance, security, remote sensing, environmental monitoring, and many others [1, 5–8] (Figure 11.1).

11.2 Automatic Detection of Hazardous Materials or Concealed Objects

The automatic detection of hazardous materials or concealed objects can be achieved using an RF radar system installed on a drone with a distributed antenna array. The radar system transmits continuous wave or pulse signals that change their parameters when they encounter and penetrate hazardous

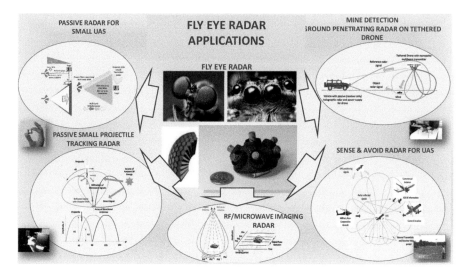

Figure 11.1 Diagram demonstrating different areas of fly eye antenna array applications, including passive radars for the detection of small UAS, super-sonic projectiles, ground penetrating radars, and sense and avoid systems. Combination of high-resolution direction finding and virtual holographic signals processing allows to apply fly eye antenna arrays as imaging radar.

materials or concealed objects. The reflected signals are then received and processed using a software-defined radio (SDR).

The SDR converts the received signals to digital form and transforms from the space–time domain to the frequency domain. In the frequency domain, the spectrum components of the signals contain valuable information about the hazardous material or concealed objects, which can be considered as a spectrum signature. These spectrum signatures can be compared with a library of saved spectrums in the processor.

By comparing the spectrum signatures of the received signals with the spectrums in the library, the radar system can automatically detect and identify hazardous materials or concealed objects. The radar system can be installed on medium- or even small-sized drones, and the use of a distributed antenna array allows for improved coverage and detection capabilities. To adapt to different environments, the radar system incorporates automatic adjustable gain in each antenna, enabling its usage in urban or mountain areas where signal conditions may vary (Figure 11.2). The effectiveness and accuracy of the automatic detection process may depend on various factors such as the design of the radar system, the quality of the spectrum library, and the specific characteristics of the hazardous materials or concealed objects being targeted.

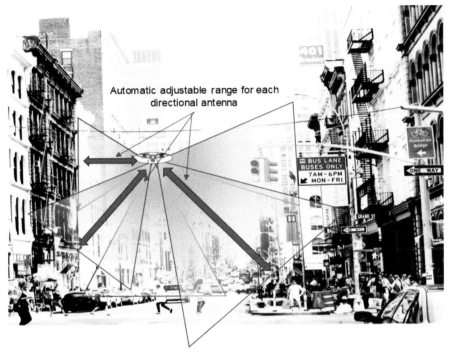

Figure 11.2 Radar for automatic detection of hazardous materials or concealed objects can be installed on medium- or even small-sized drones with distributed antenna arrays. Automatic adjustable gain in each antenna allows to use the radar in urban or mountain areas.

The use of overlap antenna patterns in multiple directions offers several advantages in radar systems:

- Simultaneous coverage of a wider area: By utilizing overlap antenna patterns, the radar system can cover a larger observation area simultaneously. This enables a more comprehensive and efficient scanning of the environment, increasing the chances of detecting hazardous materials or concealed objects.

- More information for reliable automatic recognition: The overlap antenna patterns provide multiple perspectives and angles of observation. This allows for the collection of more information about the material or object being detected, enhancing the accuracy and reliability of automatic recognition algorithms. The additional information helps in distinguishing between different materials or objects and reducing false positives.

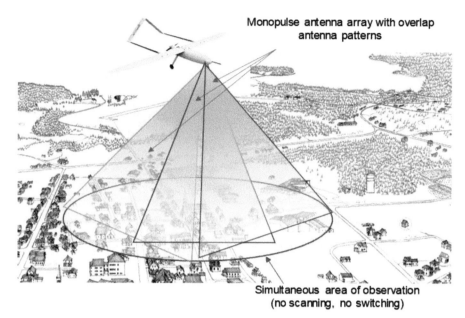

Figure 11.3   Fly eye radar with overlap antenna patterns simultaneously covering a wide area of observation and can automatically detect hazardous materials or concealed objects and their position. Overlap antenna patterns provide higher directional accuracy and increase the amount of information received about objects.

- Improved directional accuracy: The overlap antenna patterns, combined with advanced signal processing techniques such as monopulse, enhance the directional accuracy of the radar system. This means that the system can better determine the position and location of detected objects, providing more precise information about their coordinates.

- Increased image resolution: In imaging radar applications, the use of overlap antenna patterns contributes to a higher image resolution. By capturing signals from different directions, the radar system can gather more detailed information about the target, resulting in sharper and clearer images. This is particularly beneficial when identifying fine details or small objects.

The utilization of overlap antenna patterns in a fly eye radar system offers significant advantages for the automatic detection of hazardous materials or concealed objects. It enhances the system's coverage, improves directional accuracy, increases received information, and enhances image resolution, all contributing to more reliable and effective detection capabilities (Figure 11.3).

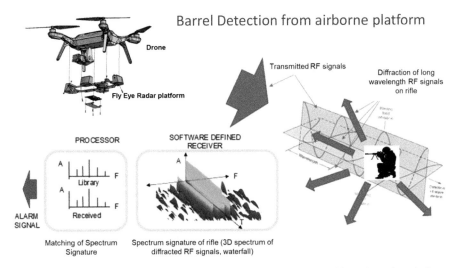

Figure 11.4 Signal processing diagram for the detection and recognition of gun barrels from airborne tactical fly eye radar platform.

The airborne fly eye radar platform can be utilized for the detection and recognition of gun barrels on tactical distances. By distributing directional antennas around the perimeter of the radar platform attached to a drone, the system can effectively cover a wide area. The signal processing diagram (Figure 11.4) for gun barrel detection application comprises the following:

1. Transmitted RF signals: Long wavelength RF signals are transmitted to the wide-covered area where potential gun barrels may be present.

2. Reflected signals: When the RF signals encounter a gun barrel, they are reflected back toward the radar platform. These reflected signals comprise resonance and diffracting components that correspond to the size and position of the gun barrel.

3. Monopulse processing: The reflected signals from the overlap antennas are subjected to monopulse processing. This technique allows for fast simultaneous high-accuracy direction finding and can determine the position of the gun barrel based on the received signals.

4. Spectrum analysis: The received signals are then processed using spectrum analysis techniques. The signals are transformed from the space–time domain to the frequency domain, allowing for a detailed analysis of their spectral characteristics.

5.   Spectrum signature comparison: The frequency–time domain data is presented as a color-coded waterfall display, showing the spectrum signature of the gun barrel. The signals contain diffraction components that correspond to specific gun barrel models.

6.   Filtering and recognition: The spectrum signatures are filtered and compared with a pre-recorded library of known gun barrel signatures. By comparing the detected spectrum signatures with the library, the system can automatically recognize and identify the type of gun barrel present.

By utilizing the fly eye radar platform and applying this signal processing diagram, it becomes possible to detect and recognize gun barrels at tactical distances. The combination of directional antennas, monopulse processing, and spectrum analysis techniques allows for accurate detection, localization, and identification of gun barrels, providing valuable information for situational awareness and threat assessment.

11.3 Mine Detection Ground Penetration Radar

The handheld mine detection radar utilizes the advantages of long wavelength RF signals for effective detection and recognition of buried objects (Figure 11.5). Here is a description of the block diagram and prototype of the tested handheld mine detection radar:

1.   Transmitted RF signals: The radar system transmits RF signals with long wavelengths, such as 1500 or 900 MHz. These signals are capable of penetrating the ground to a significant depth, with 1500-MHz signals reaching approximately 0.5 m and 900-MHz signals reaching up to 1 m.

2.   Ground penetration: The long wavelength RF signals can penetrate the ground and interact with buried objects, including mines. This allows for the detection of objects that are concealed beneath the surface.

3.   Near-field diffraction: When the RF signals encounter objects that are smaller than their wavelength, diffraction occurs in the near-field region. This phenomenon creates near-field frequency components that are proportional to the integral dielectric coefficient and size of the objects. This means that the diffraction components carry information about the objects' material properties, regardless of their position.

4.   Near-field detection: The radar system operates within the near-field region, which is limited to a distance of approximately 5–10

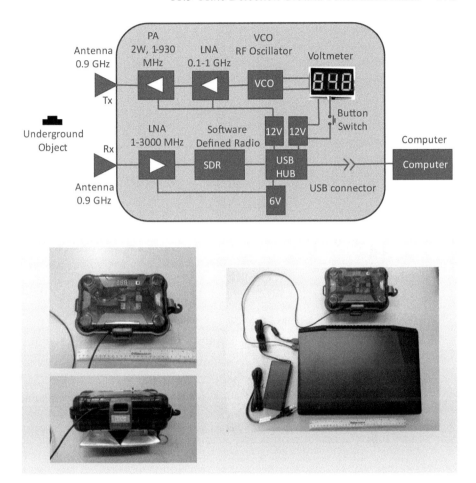

Figure 11.5 Block diagram and prototype of tested handheld mine detection radars.

wavelengths depending on the RF transferring medium. For example, with a frequency of 1 GHz corresponding to a wavelength of approximately 30 cm, the diffracted components can be detected at distances of 1.5–3 m. This range is suitable for handheld or low-altitude drone-based ground penetrating radar applications.

5. Spectrum signature recognition: The received signals, containing the diffracted components, are processed and analyzed using spectrum signature recognition techniques. Spectrum signatures corresponding to different objects, including mines, have been recorded and stored in a library.

6. Object recognition: The detected spectrum signatures are compared with the pre-recorded library of known signatures. By matching the detected signatures with the library, the system can recognize and identify the presence of objects, such as mines, based on their unique frequency components.

The handheld mine detection radar, based on the application of spectrum signatures for object recognition, offers a portable and effective solution for detecting buried mines. By utilizing the advantages of long wavelength RF signals and near-field diffraction, the system can penetrate the ground, detect objects, and identify them based on their frequency characteristics. This technology has the potential to improve mine detection and reduce the risks associated with mine clearance operations.

The fly eye ground penetrating radar has been successfully designed and tested for the detection of non-metal objects hidden underground as well as objects hidden inside laptops that contain metal, wires, and electronic components. Here are some key points regarding the testing and capabilities of the radar:

1. Detection of plastic bucket with fertilizers: In one test scenario, a plastic bucket containing a fertilizer (which can be used as a component of improvised explosive devices) was buried underground at a depth of 1 foot. The fly eye radar, with its ground penetrating capability, was able to detect and locate the hidden bucket.

2. Detection of non-metal object inside a laptop: Another test involved detecting a non-metal object (fertilizer in plastic bag) concealed inside a laptop. The laptop itself contains metal, wires, and electronic components. Despite these metallic and electronic elements, the fly eye radar was still able to identify the non-metal object within the laptop.

3. Spectrum signature recognition: The radar system employs spectrum signature recognition techniques to identify and differentiate various objects. By analyzing the received signals and comparing them with pre-recorded spectrum signatures in its library, the system can recognize and classify different objects, even if they are concealed or surrounded by other materials.

Figure 11.6 illustrates the fly eye radar being tested for the detection and recognition of non-metal underground objects as well as objects hidden inside laptops. The radar's ability to penetrate the ground and analyze the spectrum signatures of detected objects enables it to identify non-metal

Figure 11.6 Fly eye radar tested for detection and recognition of a non-metal underground object hidden underground or even inside a laptop, comprising metal, wire, and electronic components.

items, even in complex scenarios involving electronic components and metal structures.

This application of the fly eye radar demonstrates its potential in security-related tasks, such as detecting concealed objects, including improvised explosive device components, and identifying potential threats. By utilizing its ground penetrating capabilities and spectrum signature recognition, the radar system offers a valuable tool for enhanced security and threat detection purposes.

In the operation of the fly eye ground penetrating radar, long wavelength RF signals are transmitted in the direction of the ground. The antenna array, which is moving above the ground, receives the reflected signals. These received signals are then digitized and transferred to the frequency domain using software-defined radio technology. The spectrum of the received signals is displayed in the time–space domain as a waterfall plot.

As the antenna array moves slowly in one direction above the ground, the spectrum components corresponding to the background show slight changes. These changes are due to variations in the properties of the ground itself. However, when the beam reaches an object, the spectrum components corresponding to that object undergo noticeable changes. Some frequency components may decrease or even disappear, while others may increase in magnitude.

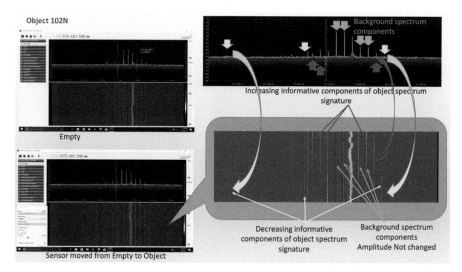

Figure 11.7 Spectrum of received signal consisting of frequency components corresponding to background and object, which can be filtered and automatically recognized by comparing with recorded a priori recorded objects library.

Figure 11.7 illustrates the spectrum of the received signal, which contains frequency components corresponding to both the background and the detected object. By filtering and analyzing the spectrum, it becomes possible to automatically recognize and identify the object by comparing it with previously recorded frequency spectrums stored in an object library.

To enable automatic detection, the radar system can be equipped with a specialized program that records and subtracts the background and/or object frequency spectrums during field tests or detection operations. This allows for the identification of objects based on their unique spectrum components and enhances the radar's ability to differentiate between the background and potential targets.

The use of the fly eye ground penetrating radar, with its ability to analyze the spectrum of received signals, provides a valuable tool for detecting and recognizing objects hidden underground or concealed within other materials. By comparing the recorded spectrum with known object signatures, the radar system can automatically identify and classify detected objects, offering significant advantages in various applications, such as mine detection, security screening, and archaeological surveys.

The fly eye ground penetrating radar offers several advantages that enhance its performance and effectiveness in detecting and recognizing objects. First, it provides simultaneous wide area coverage without the need

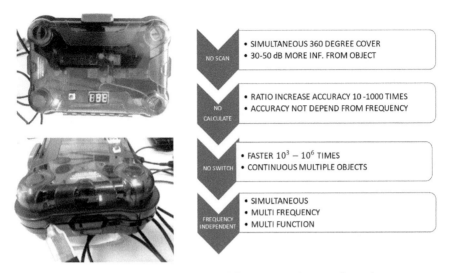

Figure 11.8 Advances of handheld fly eye ground penetrating radar.

for scanning beams. This means that the radar can gather information from a wide area without losing time on beam scanning, resulting in faster data acquisition.

The use of staring beams in conjunction with a multi-axis directional antenna array allows for the gathering of information about an object from multiple directions. This multi-axis approach enables the radar to collect a comprehensive set of data, leading to more reliable object recognition and higher accuracy in object positioning [2–4].

By utilizing overlap antenna patterns, the radar system can determine the position of an object quickly and accurately. The ratio of signals in the overlap antenna patterns provides a reliable and efficient method for object detection and recognition. Additionally, one antenna in the overlap antenna set can be used as a reference for suppressing noise, clutter, or interference signals, further enhancing the radar's performance.

Furthermore, the use of multi-channel simultaneous signal processing significantly reduces processing time. While the time required for object detection can be as short as a few microseconds, typical computer calculations take milliseconds, which is significantly longer. However, with multi-channel simultaneous signal processing, the processing time can be reduced even further, enabling faster and more efficient data analysis. Figure 11.8 illustrates the advantages of the handheld fly eye ground penetrating radar, highlighting its ability to provide fast and accurate object detection and recognition. The

Figure 11.9 Sample of application of handheld explosives/drugs detection radar for airport and stadium security.

combination of wide area coverage, high-accuracy positioning, multi-axis directional antenna array, and efficient signal processing contributes to the overall effectiveness of the radar system.

The handheld explosives/drugs detection radar, as depicted in Figure 11.9, offers a practical and effective solution for enhancing security in locations such as airports and stadiums. This radar system is designed to detect and identify hazardous materials and objects while providing ease of use and reliable performance.

The radar operates at low transmitting RF power levels, similar to that of a cellphone, with a maximum power of 50 mW. Despite the low power, it is capable of detecting hazardous materials and objects at distances of up to 10–20 meters. The long wavelength RF signals employed by the radar system enable good penetration inside passengers' belongings, ensuring reliable detection even if the hazardous objects are concealed within metal boxes, containers, or even inside laptops, as shown in Figure 11.6. The directional antenna array utilized in this radar system allows for the determination of the direction to the hazard object position with a high degree of accuracy, typically within a few degrees. This capability enhances the ability to locate and respond to potential threats effectively.

One notable advantage of this handheld radar system is its quick deployment and operational readiness. It requires minimal preparation time, allowing for immediate use in security scenarios. Radar can be applied

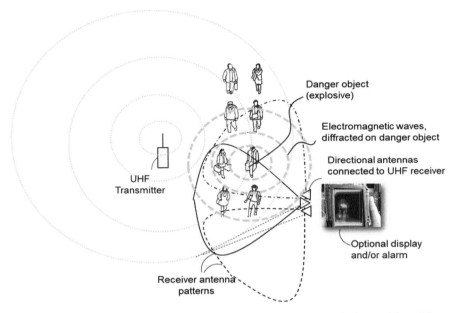

Figure 11.10 Hidden security system for automatic detection and pointing position of dangerous objects.

in airport and stadium security settings. Its compact and portable design, combined with its reliable detection capabilities, make it a valuable tool for enhancing security measures and ensuring the safety of individuals in public spaces.

The application of a directional antenna array in a portable handheld bi-static radar system offers high accuracy in pointing to hazard objects. This system utilizes a low-power ultra-high frequency (UHF) transmitter, which provides sufficient signal power for detecting hidden objects at distances of up to 10–20 m. The UHF signals have good penetration capabilities, allowing for reliable detection even when the objects are concealed within belongings.

The power level of the UHF signals used in this radar system is typically around 10–20 mW, which is considered safe for passengers and can be compared to the signals transmitted by a local broadcasting radio station. This ensures that the radar system does not pose any health hazards or interference concerns.

Figure 11.10 illustrates a hidden security system that incorporates the portable handheld bi-static radar system for automatic detection and accurate pointing of dangerous objects. The system is designed to enhance security

Figure 11.11   Sample of application of handheld human detection ground penetration radar.

measures and provide a reliable means of identifying potential threats in various settings.

By utilizing directional antenna arrays and low-power UHF signals, this hidden security system offers an effective solution for detecting and locating hazardous objects without compromising the safety and well-being of individuals in the vicinity.

Fly eye radar systems can indeed be versatile and adaptable for various applications by employing different antenna arrays, operating frequencies, and customizable processing codes. One such application is the design of a ground penetration radar, specifically for human detection, as depicted in Figure 11.11.

In this scenario, the signal processing code for the radar system typically comprises two main components. The first component is focused on detecting the spectrum signature associated with humans, enabling the radar to identify the presence of a person in the scanned area. The second component is dedicated to capturing and analyzing breath and heart movement pulses, which further enhances the reliability of human detection.

By utilizing a handheld human detection ground penetration radar, security personnel or rescue teams can effectively locate individuals who may be trapped underground or hidden from view. The radar system's ability to penetrate the ground and accurately detect human presence can be instrumental in various situations, such as search and rescue operations or security screenings.

The versatility of fly eye radar systems allows for the development of specialized configurations tailored to specific applications, ensuring efficient and reliable detection capabilities in diverse scenarios.

11.4 Microwave Bomb-detecting Imaging System

The microwave bomb-detecting imaging system based on fly eye radar technology offers several advantages over the existing X-ray technology. Unlike X-ray systems, which have limited range and pose potential radiation hazards, the low-power monopulse microwave imaging system provides enhanced penetration capabilities through camouflage and foliage.

By utilizing low microwave frequencies, the radar system achieves increased detection probability, classification accuracy, and precision imaging. The transmitter power can be significantly reduced while still maintaining effective exploration capabilities, making it a safer alternative. The system operates based on radar technology, which allows for wide field of view capabilities and improved imaging resolution.

The microwave bomb-detecting imaging system can be implemented as a handheld device, combining the transceivers with displays for convenient operation, as depicted in Figure 11.11(a). Alternatively, it can be configured as a remotely separated bi-static radar system, as shown in Figure 11.11(b). In the latter configuration, a reference antenna is employed to record real-time digital holograms, enabling 3D image reconstruction for enhanced imaging capabilities.

Overall, the use of fly eye radar technology in the microwave bomb-detecting imaging system provides a safer, more versatile, and efficient solution for detecting and imaging explosive devices, with improved penetration through various materials and enhanced imaging capabilities.

The proposed microwave imaging system, depicted in Figures 11.11 and 11.12, builds upon the concept of the passive monopulse direction finder introduced by Stephen E. Lipsky in the 1980s [4]. Monopulse involves receiving a signal simultaneously in a pair or set of antennas that cover the same field of view and then comparing the ratios of these signals. The angle information obtained through monopulse is always in the form of a ratio, which remains unaffected by signal power, common noise, or modulation.

The monopulse microwave imaging system, based on the fly eye radar technology, can operate in passive, monostatic, or bi-static modes. The system utilizes a monopulse antenna array with wide field of view capabilities to create and record real-time digital holograms, as shown in Figure 11.13.

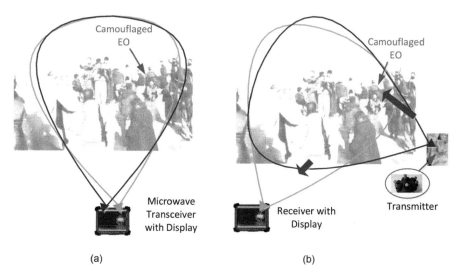

(a)   (b)

Figure 11.12   Handheld microwave bomb-detection and imaging system based on the mono-pulse antenna array offering wide field of view capabilities. Reference antenna is used to record real-time digital holograms and perform 3D image reconstruction.

The digital hologram is sampled directly through the antenna array, and the recorded microwave signals are processed in the time/phase domain.

A reference antenna is employed to provide high-accuracy phase and amplitude information. The image is then reconstructed from the phase-shifted interferograms, allowing for high-resolution imaging independent of beamwidth and diffraction limitations. The resolution of the image is determined by the sampling period and the digital phase resolution.

Proposed monopulse microwave imaging system is integrating the fly eye radar technology and offers the capability to create high-resolution images by processing real-time recorded microwave signals. By leveraging the advantages of monopulse and digital holography, the system provides accurate imaging independent of beamwidth and diffraction limitations, resulting in improved image quality and resolution.

In the estimation of the sampling frequency for holographic image reconstruction, is considering the speed of electromagnetic waves through air, which is approximately 1 foot per nanosecond. Based on this, the required sampling frequency to achieve a certain resolution can be determined.

For a resolution of 1 foot, we would need a sampling frequency of 1 GHz, as each sample would correspond to a nanosecond interval. Similarly, for a resolution of 3 cm, we would require samples to be collected at 100 picosecond intervals, corresponding to a sampling frequency of 10 GHz.

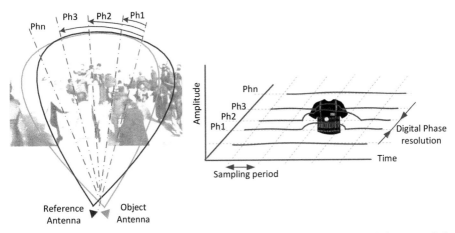

Figure 11.13 Reconstructing the digital hologram image by processing real-time recorded microwave signals in time/phase domain. Reference antenna provides high accuracy phase and amplitude information. Image is reconstructing from phase shifted interferograms. Image resolution is independent of beamwidth and diffraction limitation constraints, but rather determined by sampling period and digital phase resolution.

The Nyquist–Shannon sampling theorem states that in order to obtain an accurate representation of a waveform, the sampling frequency should be greater than twice the maximum frequency present in the signal. In the case of analog-to-digital converters (ADCs), this means the clock frequency (F_s) should be at least twice the analog input frequency (F_{in}) for faithful representation.

Figure 11.14 illustrates the concept of undersampling, where the sampling frequency (F_s) is lower than twice the analog input frequency (F_{in}). Undersampling allows for a reduced ADC sampling frequency while still capturing an aliased representation of the waveform. However, it should be noted that undersampling introduces the possibility of aliasing and distortion in the reconstructed signal. The choice of sampling frequency depends on the desired resolution and the Nyquist–Shannon sampling theorem. Higher sampling frequencies are required for finer resolution, but undersampling can be used under certain conditions to reduce the ADC sampling frequency, although it comes with the risk of introducing aliasing effects.

Figure 11.14 presents an example of ADC undersampling (Nyquist < F_{in}), as less than two samples are captured per period. To state the last sentence in equation form, when $F_{in} > F_s/2$, the ADC is undersampling the input waveform. Undersampling allows for an aliased representation of the waveform to be captured [11–13].

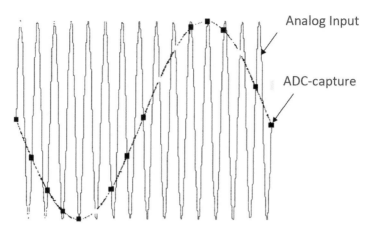

Figure 11.14   Received radar signal waveform can also be restored with adequate resolution by undersampling $F_{in} > F_s/2$. Undersampling will allow for reduced ADC sampling frequencies.

To extend high frequency limit and provide high sensitivity over the broadband of frequencies, Stephen E. Lipsky proposed to integrate detector and mixing elements with antenna [12]. Multiple angular offset directional antennas can be integrated to fly eye antenna array are coupled with front-end circuits and separately connected to a radar seeking processor through a digital interface. Integration of antennas with front-end circuits allows to exclude waveguide lines, which limits the system bandwidth and creates frequency-dependent phase errors. Digitizing of received signals in close proximity to antennas dramatically decreases phase errors associated with waveguides. Image resolution, in this case, will be determined by the accuracy of amplitude and phase measurements, which are a function of digital processing time accuracy and sampling frequency.

Figure 11.15 presents an algorithm for creating high-resolution 3D images from a digital hologram. The digital information captured by multiple object antennas can be transferred to the image processor using fiber optic cables. The algorithm combines phase-shifted interferograms to generate 3D images. To improve image resolution, multi-frequency phase domain processing techniques are employed.

In Figure 11.16, the application of Kalman filters and SMART (spatially matched adaptive radar tomography) processing is demonstrated in the context of a fly eye radar system. This processing enables the estimation of velocity vectors, which consist of both speed and direction of motion (a). By applying eigen-harmonic decomposition approximation, the image can be

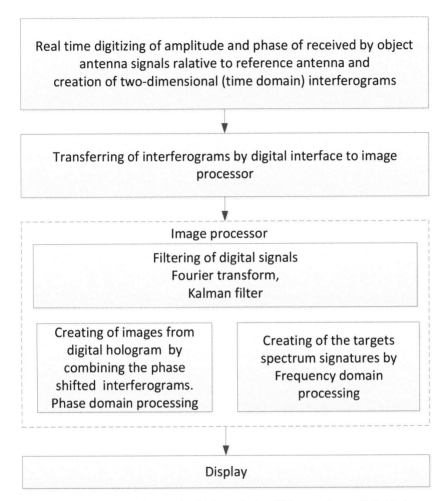

Figure 11.15 Algorithm for creating high-resolution 3D images from a digital hologram.

enhanced, particularly in the presence of debris or cluttered backgrounds. The presented figures highlight advanced techniques and algorithms used in radar systems, such as digital holography, multi-frequency phase domain processing, Kalman filters, and SMART processing, to achieve high-resolution imaging and improve the accuracy of velocity estimation.

11.5 Improving of SNR by Coherent Integration

There are a few accuracy parameters that need to be taken into account to determine approximate reconstructed image resolution. Range accuracy and

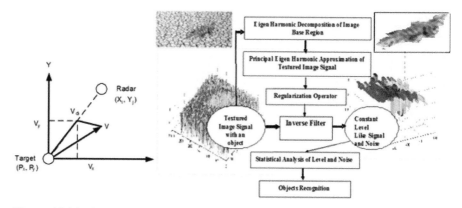

Figure 11.16   Application of Kalman filters and SMART processing of received signals in the fly eye radar can provide velocity vectors consisting of speed and direction of motion (a). Eigen harmonic decomposition approximation allows to enhance image on debris background.

range resolution are parameters usually used in radar systems. Range accuracy, δR, is related to range resolution, ΔR, as follows [10]:

$$\delta R \cong \frac{\Delta R}{\sqrt{2SNR}} [\text{m}] \text{ for SNR} \gg 1. \qquad (11.1)$$

Accurate range measurement is different (though related) from range resolution.

The ability to accurately extract the round-trip time of flight depends on the range resolution, ΔR, and on the signal-to-noise ratio, SNR. It can be shown that range accuracy, δR, is related to bandwidth, B, and SNR as

$$\delta R \cong \frac{c}{2B\sqrt{2SNR}} [\text{m}]. \qquad (11.2)$$

For example, for SNR $\gg$ 1, consider a radar with $B = 2000$ MHz and an SNR of 20 (13 dB).

The achievable range resolution, ΔR, is 0.5 m and the achievable range accuracy, δR, is 1 cm.

In general, the uncertainty associated with any measurement is related to the measurement resolution. For measurement with resolution ΔM, the accuracy, δM, is

$$\delta M \cong \frac{\Delta M}{\sqrt{2SNR}} [\text{m}] \text{ for SNR} \gg 1. \qquad (11.3)$$

High pointing accuracies of phase measurements provide high accuracy imaging. If antenna lobes are overlapping or closely spaced, a monopulse antenna array can produce a high degree of pointing accuracy within the beam, adding to the natural accuracy of the conical scanning system. Classical conical scanning systems generate pointing accuracy on the order of 0.1°, whereas monopulse radars generally improve this by a factor of 10, and advanced tracking radars like the AN/FPS-16 are accurate to 0.006° [14]. High resolution measurement data allows for the reconstruction of high-resolution images. By coherently adding echo signal energy from consecutive pulses where effectively increasing the illumination energy. This may be thought of as increasing the transmitted power, P_t. The radar transmitter has peak output power and a pulse duration, and the transmit pulse energy is: $P_t \cdot \tau$. Coherently integrating echoes from 10 pulses ($N_{coh} = 10$) produce an SNR equivalent to the case where P_t is 10 times greater.

Alternatively, coherent integration permits a reduction of the transmit pulse power P_t equivalent to the N_{coh} while retaining a constant SNR [10]:

$$\sum_{n=1}^{N_{coh}} \quad \boxed{Tx_n} \qquad \boxed{Tx} \quad N_{coh} P_t \tag{11.4}$$

Coherent integration provides significant SNR improvement. To realize the full benefits of coherent integration the following assumptions must be satisfied:

- Noise must be uncorrelated pulse to pulse. Signal phase varies less than 90° over integration interval.

- The second assumption limits the integration interval for cases involving object moving relative to the radar.

- Coherent integration can be used if phase variation is removed.

  Processes involved include range migration and *focusing*.
  For a 2.25-kHz PRF, $N_{coh} = 100{,}000$ or 50 dB of SNR improvement.
  The directional monopulse antenna array consists of angular shifted antennas and has overlapping antenna patterns. One antenna can be used as a reference for high-accuracy amplitude/phase measurements. Each antenna is integrated with a front-end digitizer circuit and is connected to an image processor through a digital interface. This allows for the storage of amplitude

and phase components of received signals as a digital hologram with high accuracy in the ultra-wide frequency band.

- Proposed microwave bomb-detection and imaging system will provide all weather high-resolution images in smoke and dust conditions with enhanced penetration of camouflage, foliage, and underground devices up to 0.5 m below the surface.

- Digitized signals in antenna module provide phase/frequency indepen-dence. As a result, the system can be ultra-wide band, multi-band, or step frequency modulated. Fourier transform of wide band reflected signal provides reliable identification and classification of EO, UXO and IEDs.

- Image resolution is determined by the processor and sampling fre-quency and is not limited by electromagnetic wave diffraction.

- The system can be set up and deployed in moments consisting of a few seconds.

- The system will be controlled with open source software and can be operated wirelessly from a laptop.

- The system will provide real-time imaging of objects without the need for safety zones as the transmitting microwave power is minimal; this is physically comparable to common cell phone operation.

- Optionally, the proposed system can be used for camouflaged human detection in foliage or ground buried mine detection.

11.6 Mine Detection with Tethered Drone

Figures 11.17 and 11.18 illustrate the concept of bi-static (optional monostatic) ground penetrating radar (GPR) for mine detection using a tethered drone. In this configuration, a low-frequency GPR transmitter is positioned on a small drone that is tethered to the ground. The low-frequency waves are ideal for ground penetration and can achieve a depth of up to 2 feet and a distance of up to 60 feet with a transmitter power of just a few hundred milliwatts.

The receiver system utilizes a multibeam monopulse radar with an array of angle-shifted directional antennas. These antennas provide high-accuracy and high-resolution measurements by utilizing a reference beam. The sig-nals received by each directional antenna are directly digitized relative to the processor's reference signals, allowing for the real-time recording of a

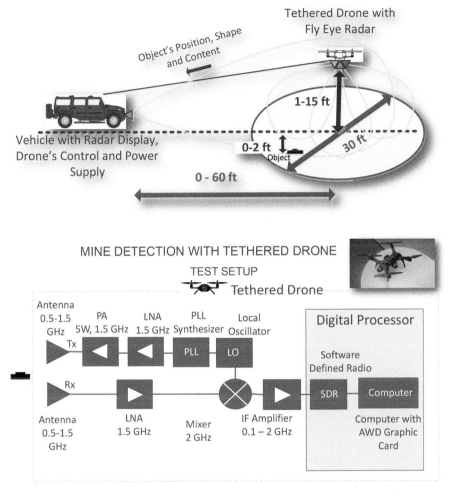

Figure 11.17 Mine detection with a tethered drone diagram and schematic.

digital hologram that contains both amplitude and phase information about underground targets.

The resolution of the digital hologram and the corresponding image resolution are determined by the sampling frequency of the digitizer, rather than the radar beamwidth. By employing high-speed sampling and a high-accuracy processor clock, the system can achieve high-resolution images even when using low-frequency radar waves.

The holographic digital phase/time domain processing of the received signals enables the restoration of images of the detected objects. Additionally,

How deep can GPR go into ground?:

It depend upon two condition:

❏ The type of soil or rock in the GPR survey area.

❏ The frequency of the antenna used.

➢ Low frequency systems are more penetrating but data resolution is lower.

➢ High frequency systems have limited penetration but offer a much higher resolution.

Antenna Frequency	Maximum Penetration Depth	Appropriate Application
1500 MHz	0.5 m	Rebar mapping and concrete evaluation.
900 MHz	1 m	Pipe and void detection or assessing concrete thickness.
400 MHz	4 m	Utility surveys, pavement evaluation, storage tank detection and assessing structural integrity
270 MHz	6m	Utility surveys, geology and archaeology

Figure 11.18   Mine detection from a tethered drone provided the possibility to detect mines for 8–12 hours because of using power from a car battery. An optimal radar position for mine detection provides a reliable detection of small cross-sectional objects. The application of spectrum signature provides the possibility of detecting non-metal and non-wire objects.

by performing a Fourier transform (frequency-domain processing) of the received radar signals, valuable information about the shape and material of the buried objects can be obtained. The use of a tethered drone in combination with bi-static GPR offers an effective and efficient approach for mine detection, leveraging advanced signal processing techniques and low-frequency radar waves for accurate imaging and identification of underground objects.

11.7 Conclusion

In conclusion, antenna arrays play a crucial role in various communication, radar, and surveillance systems. The development of advanced antenna array technologies, such as fly eye antenna arrays, offers numerous benefits and solutions to overcome the limitations of traditional scanning or mechanically rotating antenna systems [15–21].

Fly eye antenna arrays provide continuous coverage of wide observation areas without the need for scanning or switching beams. This enables faster detection and tracking of targets, making them well-suited for applications like drone detection or tactical systems. The integration of signals from multiple antennas in the fly eye array dramatically increases radar range and information about targets, enhancing sensitivity and detection capabilities.

While fly eye antenna arrays may have a smaller range compared to narrow beamwidth scanning systems, they excel in wider area coverage and faster detection times. This makes them ideal for scenarios where a comprehensive view of the surrounding environment is required, such as in all-space communication or surveillance systems.

Additionally, the use of directional antennas in fly eye arrays allows for compact module arrangements, distributed placement, and flexibility in different platforms. The small size of the antenna elements and the ability to position them close together enables effective utilization of limited space, even for long wavelength RF systems.

Moreover, advancements in signal processing techniques, diversity signals, modulation, and intelligent processing further enhance the performance of antenna arrays. These techniques help increase detection reliability, recognize and classify signals or targets, and improve sensitivity and range.

The fly eye antenna array technology offers promising solutions for achieving maximum coverage, sensitivity, range, and line budget optimization in various communication, radar, and surveillance applications. Continued research and development in this field will drive further advancements and enable the realization of more efficient and effective antenna systems for future technologies.

References

[1] P. A. Molchanov, A. Gorwara, "Fly Eye radar concept". IR 2017. International Radar Symposium, Prague, July 2017.

[2] S. A. Harman, "A Comparison of staring radars with scanning radars for UAV detection" Proc. of the 12th European Radar Conference, Paris, France, Sept 2015.

[3] S. A. Harman, "Holographic radar development", Microwave Journal, Vol. 2, February, 2021, www.microwavejournal.com/articles/35410-holographic-radar-development.

[4] S. E. Lipsky, "Microwave passive direction finder", ciTech Publishing Inc. Raleigh, NC 27613, 2004.

[5] P. Molchanov "Tactical radar system for detection of hypersonic missiles and UAS", patent appl. 17/651,800 02/02/2022.

[6] P. Molchanov "Passive radar system for detection of low profile low altitude targets", US patent appl. 17/971,582, 10/22/2022.

[7] P. Molchanov "Multi-beam multi-band antenna array module" US patent appl. 17/971,616, 10/23/2022.

[8] P. Molchanov "Multi-beam multi-band protected communication system" patent appl. 17/740,581, 05/10/2022.

[9] Houngsun Yang, Seungheon Kim, Soonyoung Chun, "Bearing accuracy improvement of the amplitude comparison finding equipment by analyzing the error". *Int. Journal of Communication Networks and Information Security.* Vol. 7, No.2, Aug. (2015).

[10] Chris Allen, "Radar Measurements", Course website URL <people. eecs.ku.edu/~callen/725/EECS725.htm>

[11] Texas Instruments, Application Report, SLAA510–January 2011.

[12] US patent 5,134,403 Kenneth Rush, Hewlett-Packard Co. High speed sampling and digitizing system requiring no hold circuit, Jul. 28, 1992.

[13] European patent application PCT/JP88/00131, 0 371 133 A1. Fujisaka Takahiko, Ohashi Yoshimasa, Kondo Mithimasa, Mitsubishi Denki Kabashiki Kaisha, JP, Holographic radar. 10.02.1988.

[14] Radar Set - Type: AN/FPS-16. US Air Force TM-11-487C-1, Volume 1, MIL-HDBK-162A. 1 5 December 1965

[15] A. Gorwara, P. Molchanov, O. Asmolova, "Doppler micro sense and avoid radar", 9647-6, *Security+Defense 2015, Toulouse, France,* 21–24 September 2015, <http://pmi-rf.com/documents/DopplerMicroSenseand AvoidRadarPaper.pdf >

[16] P. Molchanov, O. Asmolova, "Sense and Avoid Radar for Micro-/Nano Robots" *(Invited Paper), Security+Defense Conference, Amsterdam, September 24,* 2014, <http://spie.org/Publications/Proceedings/Paper/ 10.1117/12.2071366>

[17] P. Molchanov, "All Digital Radar Architecture" Paper 9248-11, *Security+Defense Conference, Amsterdam, September 25,* 2014, <http:// spie.org/Publications/Proceedings/Paper/10.1117/12.2060249>

[18] P. Molchanov, V. Contarino, "Directional antenna array for communications, control and data link protection." *2013 Defense + Security. Session 9, Communication, Control, and Enabling Technologies, Paper 8711-31,* April 30, 2013, <http://spie.org/Publications/Proceedings/Paper/10.1117/ 12.2015607>

[19] V. Contarino, P. Molchanov, R. Healing, O. Asmolova, "UAV Radar System of Low observable Targets." *Conf. Unmanned Systems, Canada. Nova Scotia, Canada*, K1P1B1, Nov. 7–10, 2011.

[20] P. Molchanov, V. Contarino, "New Distributed Radar Technology based on UAV or UGV Application", *Defense +Security. Session 6, MIMO Radar*, Paper 8714-27, April 30, 2013.

[21] U.S. Patent and Trademark Office (USPTO) application titled "Multi-beam Antenna Array for Protecting GPS Receivers from Jamming and Spoofing Signals" (reference USPTO EFS-Web Receipt 13562313 dated 31 July 20, 2012, (http://www.google.com/patents/US20140035783).

Index

About the Author

Pavlo A. Molchanov (Senior Member, SPIE, IEEE) is a highly accomplished professional with an impressive educational background and extensive experience in the field of engineering. He received the M.S. degree from the Radio Technical and Electronic Devices Department, Kiev Polytechnical Institute, Ukraine, and the Ph.D. degree from the Radio Electronics Department of Moscow Aviation Institute (MAI), Moscow, USSR, in 1971 and 1980, respectively. He has designed radars and antenna systems for low-altitude missiles. He has been a US citizen since 2003. Dr. Molchanov has made significant contributions to the field. He has served as a Principal Investigator in various projects funded by the U.S. Air Force Small Business Innovation Research (AF SBIR) program and the Office of Naval Research (ONR). In these roles, he designed LIDAR/RADAR systems, protected GPS systems, and developed sense and avoid antenna arrays for U.S. Navy Ships and Naval Aviation.

Dr. Molchanov's expertise is well-documented through his extensive publication record, which includes 240 publications, 4 textbooks, and 20 US patents. His work has significantly contributed to the advancement of engineering and technology in the field of antenna arrays and related areas.